D0934365

INTERSTATE

Mark H. Rose is a research associate in the history of technology and urban affairs at The Franklin Institute in Philadelphia. He has taught American social, business, and urban history and served as associate editor of *American Studies*. He has published articles in urban and business history and is also coeditor of a series in technology and urban environmental formation.

MARK H. ROSE

INTERSTATE

Express Highway Politics
1941-1956

THE REGENTS PRESS OF KANSAS
Lawrence

Printed in the United States of America

Library of Congress Cataloging in Publication Data
Rose, Mark H 1942-
Interstate.
Bibliography: p.
Includes index.
1. Express highways—United States—History.
2. Express highways—Law and legislation—United States—History. 3. Transportation and state—United States—History. I. Title.
HE355.R675 388.1'2 78-14940
ISBN 0-7006-0186-4

TO

SWEETHEART

Contents

Preface

Transportation is a key element in the social and economic organization of any industrially developed society. As the construction of railroads in America influenced the location of cities, so the development of streetcar lines helped to shape them. Both forms of transportation contributed to employment opportunities and economic growth, indeed to the course of the industrial revolution. Obviously, highways, too, are a part of the transport system. Throughout the twentieth century, construction and paving of several million miles of roads have facilitated migration and travel and created jobs and encouraged investment. Homeowners, farmers, and businessmen followed roads in and out of urban centers, thus reshaping cities, further restructuring social and economic networks.

Scholars, however, have paid surprisingly little attention to road promotors, to leaders of trucking firms, planners, farmers, and engineers and their competing visions of highway building, economic growth, faster traffic, and renewed cities. Social critics instead have churned out the bulk of analysis and opinion. Highway construction, as they explain it, represented the success of legislative intrigue. An unsavory alliance of politicians and lobbyists, according to this drama, imposed unneeded roads on a foolish and gullible public, in the process ruining mass transit, creating useless jobs, and destroying rustic charm. The popular view of road building appears more jaundiced. Americans, while they have always welcomed new highways, believe that road legislation served as license for realtors and contractors to raid the United States Treasury.

Development of the American highway system, especially express highways, involved more than squabbling about spoils and far more than subtle political maneuver. By 1910 or so, America was already the leading automobile culture; by the late 1920s, more than one-half of American families owned a car. As truck and auto sales mounted, owners accepted the burden of financing road construction themselves, largely from gasoline sales taxes. But engineers, however quickly they worked, failed to build and surface highways fast enough. Depression and inflation, unpredictable and expensive developments over which no one had control, slowed construction.

Different visions of redeveloped cities, of faster traffic, and of a less bumpy economy, each in turn supported by competing engineers, planners, and leaders of three presidential administrations, delayed construction further. In 1944, after many years of planning at the state and local levels, Congress authorized construction of the National System of Interstate Highways but refused to appropriate funds to build it. Not until 1956, when the press of traffic appeared overwhelming, could engineers, truckers, auto club executives, and urban and national political leaders agree to a program of accelerated road construction.

This study, then, leads in several directions. I explore the processes of building and shaping cities, most notably the efforts of professional men and of business leaders to use highways to determine the outlines of urban places. Economic growth, untroubled by inflation and downward tumbles, was a prize sought by those in both the road transport and construction fields and by government economists and political leaders. But I examine the politics of economic policy making and find that each perceived sound growth in different terms. In brief, economists planned road building as part of their own picture of the national economy; truckers, mostly local businessmen, endorsed the idea of economic growth, but objected when federal officials altered the rate of highway spending to remedy the economy. Finally, I look at the development of transportation facilities themselves, particularly as planners, economists, and businessmen hoped to use them as adjuncts to broader professional and commercial goals. Ultimately, what I present is an analysis of specialized points of view, those which flowed from participation in professional and business life, those which in turn influenced leaders to reshape business institutions and urban centers. In the roadbuilding industry, while many opted for national planning, those who spoke for autonomy, for the imposition of professional standards, and for federal funding triumphed. Ideally, my work will serve as a background for more intensive examinations of other industries, more cities, and other facets of the postwar experience, particularly for study of the social bases of competition between elites.

To acknowledge assistance, counsel, and encouragement, all through six years of research and writing, is a delight. A grant from the Department of History at The Ohio State University and two grants from the Office of Research Administration at The University of Kansas helped greatly with travel and typing costs. The Graduate School at Ohio State awarded a fellowship, allowing a year for more travel, for earlier writing, for lengthier periods in the library. Archivists and librarians went above duty. Carmen Delledonne, William E. Lind, and William Sherman at the National Archives and Records Center and Stanley W. Brown at the Washington National Records Center escorted me through many collections and turned up items on their own initiative. George C. Curtis, James Leyerzapf, Jo Ann Williamson, and Don W. Wilson, members of John E. Wickman's staff at the Dwight D. Eisenhower Library, compiled research lists, secured release of classified material, and even looked after housing arrangements for myself and family. Archivists at the Harry S. Truman Library, especially Dennis E. Bilger, Peter W. Bunce, and Philip D. Lagerquist, were equally helpful and cordial. Although my visit was a short one, members of the staff of the Franklin D. Roosevelt Library found what I needed. Ivy Parr, chief of the Records Management Division of the Department of Commerce, inconvenienced herself and aides by making space in which I could work. Richard C. Creighton, head of the Highway Division of Associated General Contractors, searched files in the basement of his Washington, D.C., headquarters and found much useful material. My collective thanks to librarians at Ohio State University and The University of Kansas who wrote several times for interlibrary loans, retrieved lost items, and stayed late to locate others. Lois E. Clark interrupted her own preparation for graduate school to type the manuscript, making important corrections, deciphering my codes.

Several people contributed to this study in special ways. Richard O. Davies, James J. Flink, Mark S. Foster, Raimund E. Goerler, David E. Green, Ellis W. Hawley, Glen E. Holt, Thomas R. Hyland, Raymond A. Mohl, John B. Rae, Gary W. Reichard, and Mary E. Young endured early versions. Colleagues at The University of Kansas—David M. Katzman, J. Robert Kent, and Lloyd L. Sponholtz—read parts of the manuscript and returned them marked in a manner I found helpful. Others at Kansas—Clifford S. Griffin, Phillip S. Paludan, and particularly John G. Clark and Donald R. McCoy—took time to discuss issues and methods, to calm an anxious assistant professor. John C. Burnham and K. Austin Kerr, both at Ohio State, made vital suggestions. Their tingling critiques and good humor taught me a great deal about historical analysis and about the spirit of professional interaction; they were patient while I tried to learn. My parents, Bertha and Albert Rose, waited thirty-five years to see one of the fruits of their subvention and hope. This small book, I trust, will in part repay their patience. Marsha Shapiro Rose, always busy with teaching, a dissertation in progress, and major responsibility for our daughter, Amy, offered sound judgments throughout, more so in fact than my reputation as a historian, such as it may be, will permit me to confess. Marsha was patient too.

1

Rebuilding America: Express Highways and Visions of Reform, 1890-1941

In highways, then, lies a new national frontier for the pessimist who thinks frontiers have disappeared. It challenges the imagination and spirit of enterprise which always have been the distinctive marks of American life. And even the gloomiest of men admit that America never ignores the challenges of a new frontier, geographical or otherwise.

Paul G. Hoffman, President,
The Studebaker Corporation, 1940

By 1960, a recorded voice promised visitors to General Motors' Futurama exhibit at the 1939 New York World's Fair, fourteen-lane express roads would accommodate "traffic at designated speeds of 50, 75, and 100 miles an hour." Spectators, six hundred at a time, rode around GM's 35,738 square foot mock-up of future America while the synchronized recording in each chair continued. Automobiles from farm and feeder roads would "join the Motorway at the same speed as cars traveling in the lane they enter," and motorists would be able to "make right and left turns at speeds up to 50 miles per hour." In urban areas, express highways would be "so routed as to displace outmoded business sections and undesirable slum areas." In cities themselves, men would construct buildings of "breath-taking architecture," leaving space for "sunshine, light and air." Great sections of farm land, "drenched in blinding sunlight" according to an observer, were under cultivation and nearly in fruit. Traffic, whether in rural or urban areas, flowed along without delays and without hazard at intersections and railroad crossings. "Who can say what new horizons lie before us . . .," asked the voice on the record, "new horizons in many fields, leading to new benefits for everyone, everywhere." By mid-May 1939, only a few weeks after the fair opened, Futurama was the most popular attraction.[1]

Actually, GM's exhibit, if fanciful, contained concepts and plans well known to engineers, business leaders, urban and regional planners, and highway-minded men. Yet between 1900 and 1939, these planners never managed to construct sufficient highway mileage, to speed-up traffic, to remodel cities and farm areas, or to put everyone to work. By the 1920s, costs for building heavy-duty roads, those designed to handle huge volumes of traffic, consumed an increasing portion of local, state, and federal road budgets. But political disputes, not budgets and designs, proved decisive in the minds of those charged with creating an adequate road system. In brief, men who judged road building according to distinct professional, commercial, and bureaucratic points of view limited their own flexibility in political arenas. As leaders—civil engineers, truck operators, planners, and farm organization executives—squabbled over limited funds, the pace of road building suffered. Legislators, themselves just as incapable of forging a coherent road program, responded to highway enthusiasts by writing construction programs to aid local and often professional interests. Rural township and county officials built farm-to-market roads and urbanites stressed attention to their own streets, but not until the 1930s did the federal government contribute to the cost of farm–market and urban highway building. State and federal road engineers, for their own part, focused on construction of main trunk roads between cities, thus ignoring crowded urban districts as well as little-used routes in the hinterlands. During the 1930s, President Franklin D. Roosevelt and his cohorts thought principally of highway building as part of a package aimed at relieving unemployment. The net result, up to the onset of war with the Axis, was the creation at each level of government of a series of complex, ambivalent, and inconsistent road programs, all of which together fostered construction of limited and often substandard highway mileage. Motorists of course just wanted more roads. Yet because they defined their own highway needs in particularistic and often localized terms, engineers built a highway system which failed to serve as expected.

Traffic increases created the framework for American highway politics. In short, auto and truck sales and the auto industry grew at a spectacular pace. In 1905, Americans registered about 78,000 vehicles; by 1910, with 458,500 motor vehicle registrations, America was already the leading auto culture in the world. In 1921 alone, Americans purchased 1.6 million vehicles, about half on credit. By the late 1920s, as registrations climbed above 23 million, a survey by the General Federation of Women's Clubs showed that 55.7 percent of American families owned a car. Of families with autos, moreover, 18 percent owned more than one. Southerners, traditionally the most culturally conservative and impoverished Americans, were nearly as excited about a new car. During the 1920s, while many Southerners worried that auto ownership would reduce church attendance and affect the morals of their children adversely, still others spoke optimistically about auto-based prosperity. Between 1920 and 1930, registrations quadrupled in Alabama and more than doubled in Georgia and South Carolina. If adjustments were made for regional differences in family size, age, and the like, argues historian James J. Flink, during the 1920s the ratio of auto ownership to population in every region would have been about the same. Even during the hard days of the Great Depression, Americans purchased new vehicles, by 1940 boosting auto registrations to 27.4 million and truck registrations to 4.8. In 1909, as demand and production increased, the auto industry ranked twenty-first in value of product; by 1925, auto products ranked first, and the industry led all others in costs of materials and wages.[2]

American technology, governmental structures, and basic values facilitated huge auto sales. In short, Detroit manufacturers operated in an environment in which autos were welcomed. Long before 1900, plenty of raw materials and insufficient labor had encouraged mechanized production and standardized output. After 1910 or so, because neither state nor federal governments imposed internal tariffs and costly safety standards, once large-scale production was under way middle-income Americans could purchase an inexpensive auto. An automobile, finally, allowed a vast expansion of personal choice and mobility, and provided new profit-making arenas for the owner. In 1907, a publicist predicted that an automobile would "remove the last serious obstacle to the farmer's success. It will market his surplus product, restore the value of his lands, and greatly extend the scope and pleasure of all phases of country life." Travelers on urban fixed-rail trolleys, on the other hand, often encountered traffic delays and hazards caused by tracks and wires, and all faced encounters with pickpockets, drunks, thugs, and the obnoxious. Streetcars themselves were dreadfully overcrowded. Passengers were "packed like sardines in a box, with perspiration for oil" and were forced to "hang on by the straps, like smoked hams in a corner grocery." By the early years of the twentieth century, then, both farmers and trolley riders were anxious to try their luck with a cheap if not always reliable automobile.[3]

So enthusiastic were auto and truck operators about the prospects for motoring and truck transport, beginning in 1919, they financed road building themselves. Between 1900 and 1920, local and state officials had paid roughly three-fourths of their highway construction bill from property and other taxes; federal road programs, a modest affair up to 1921, drew exclusively upon general funds. In 1919, legislatures in Oregon, New Mexico, and Colorado imposed a penny a gallon tax on gasoline sales. By 1929, every state collected a tax on gas, ranging from two to six cents a gallon, and twenty-one states had dropped property taxes as a source of funds to construct main trunk roads. Between 1930 and 1940, as the economy and tax receipts in other areas fell, average state gas tax rates increased two-thirds of a penny and income soared from $494 million to $870 million. In 1932, to make up for declining revenues, Congress imposed a one cent per gallon levy. Federal tax income, however, was ploughed into general receipts, not set aside for road construction. But if state tax receipts were dedicated to highway building, rarely did motorists and truckers object, even when rates went up. "Motor vehicle owners," declared the author of a trucking industry and auto club publication in 1941, "are responsible for the major share of the cost of primary highways and a minor share of the cost of secondary roads."[4]

As auto and truck sales soared, motorists, government officials, and engineers turned their attention to highway building. Between 1921 and 1940, government officials at all levels spent $34.6 billion for road construction and repair; during the 1920s, in fact, road costs added up to the second largest area of governmental expense. Engineers, in return, constructed 418,000 miles, boosting the total to 3.0 million. Surfacing—defined by engineers as an application of sand, stone, gravel, soil, concrete, or asphalt—was much less costly and thus went ahead faster, jumping from 387,000 surfaced miles to 1.3 million. Limited access roads, although relatively short in length, were opened in Chicago, Los Angeles, New York City, and Pittsburgh, and in Connecticut, Massachusetts, New York, and Pennsylvania. As automobile and truck ownership increased, men active in highway affairs sketched even larger projects, occasionally including urban and cross-country expressways. Beginning in March, 1939, Thomas H. MacDonald, chief of the Bureau of Public Roads in the Department of Agriculture, was busy promoting plans to construct a 30,000-mile national expressway system, one aimed at speeding rural and urban traffic, eliminating urban and rural decay, and creating useful jobs.[5]

If Americans, especially motorists, believed that highway building itself was justified, the consensus ended there. Auto and truck sales increased faster than highway mileage. In part, competition for new mileage and for limited funds delayed construction both in urban centers and along increasingly jammed rural routes. As early as September 3, 1912, Ohio voters turned down a road bond issue. In short, those located in areas with above average roads objected to financing construction for the improvident. During the 1910s and later, demand for road building and rising costs forced engineers to follow old rights-of-way and to construct in stages, here and there finishing narrow and curving highway strips. From 1911 through 1923, for example, Delaware engineers worked fitfully on a ninety-six-mile section of routes 13 and 133; but costs doubled and design was changed from divided and multilane to two lanes totaling eighteen feet, another conventional road. Frequently, only minimum surfacing was put in place, usually enough to "get the traffic through"; few roads were constructed for increasing numbers of faster and heavier vehicles. In 1934 and 1936, members of Congress voted aid to build extensions of trunk and farm–market roads, thus within a couple of years dispersing funds across more than 80,000 additional miles. But by March 3, 1938, state and federal road engineers persuaded members of the Senate to block authorization of an $8 billion toll-superhighway network aimed at creating jobs, first because it would not serve traffic as effectively as a toll-free highway, next because creation of a construction authority threatened federal-state links. "Toll highways . . .," the head of the New York State Public Works Department wrote Public Roads Chief MacDonald on February 15, 1938, "as are proposed in this bill are utterly impracticable." As it stood, "one of the worst features of the bill is that it by-passes all existing, experienced agencies such as your Bureau and the several State departments." It was, he added, "a crackpot bill."[6]

If few could agree about highway tax and finance items, fewer still had reached agreement about the proper impact of new roads. City planners and their friends in the business community promoted highway construction, by the 1930s especially express highways, with a view toward rescuing their cities. As urbanites moved to the suburbs of deteriorating and congested cities, planners insisted that an accelerated road program would hasten traffic flow and boost morale and economic development, thus in general upgrading urban life. Highway building, in their scheme, was a form of social and economic therapy.

Beginning during the 1890s, business and professional men in Kansas City endorsed a vast program of highway construction. In

1893, George E. Kessler, a landscape architect employed by the Board of Park and Boulevard Commissioners of Kansas City, presented a proposal to board members to construct the Paseo and other highways as well as parks and fountains for the city. Boulevards a hundred feet wide, themselves flanked by parkways thirty feet across and lined with trees and walkways, were supposed to check decentralization, reduce congestion on streetcars, separate homes from factories and shops, and prove delightful places along which to shop. Kansas Citians, Kessler expected, ultimately would enjoy improved health and sounder relationships with one another. In order to promote "public health and comfort . . . and the culture and refinement of our people," resolved business leaders at a Commercial Club meeting on December 6, 1893, purchase of property and construction features as Kessler recommended should begin "as speedily as possible."[7]

By 1910, promises of efficient cities competed with city beautiful ideals in the minds and schemes of planners. Now, technicians of land-use planning calculated more variables, normally including the location of public buildings and waste disposal, transport coordination, and the like, but their overall social emphasis remained much the same. In 1914, Charles F. Puff, Jr., the author of a city plan for Newark, New Jersey, proposed that planners and engineers consider block size and transport requirements as well as the location of parks, markets, and schools in their efforts to improve housing conditions and reduce congestion. "Why lose this splendid opportunity," asked Puff, "for moulding characters?" Charles M. Robinson—professor of civic design at the University of Illinois, author of a book on street layout, and for years an enthusiastic city beautifier—perceived a direct relationship between street planning and social and health problems. "In short," he announced in 1914, "the street has a duty to perform not only in protecting the health of those who live upon it, but in also protecting that of the community." During the progressive era, highway planning made up a visual and physical component of urban revitalization and social reform.[8]

After 1920 or so, city planners focused on street and then expressway planning, neighborhood social life, and reconstitution of their central business districts. Whether they opted for a scattering of business and population centers or for increased centralization, and indeed professional planners were badly divided on the matter, almost all endorsed the proposition that construction of a coordinated road network along with other civic improvements would enhance neighborhood integrity, upgrade downtown property values, and make men and their families wealthier and better behaved too. From about the mid-1920s through the end of the 1930s, the work of the Committee on the Regional Plan of New York—a group of planners, business leaders, and technicians headed up by Thomas Adams—exercised the leading influence in urban and regional planning circles. They proposed broad programs for decentralizing and coordinating the location of manufacturing and commercial establishments and housing, in part by more sophisticated alignment of highway coordinates. "Wide radial highways with adequate connecting roads," or so claimed Adams in 1927, "will facilitate a rational degree of dispersal and closer contiguity of industry and residence." Good roads, themselves properly located, were supposed to create more viable neighborhoods, to aid downtown property values. "Between the radial lines of transportation," moreover, it was possible to plan "for the development of neighborhood districts," both for "new residential growth in the suburbs and the reconstruction of central areas."[9]

By the late 1930s, professional city planners had emerged in American society capable of emphasizing in policy discussions the importance of giving long-range, elite guidance to city development. Some focused on the aesthetics of the urban environment; others preferred to direct the distribution of population so as to channel social integration; still others opted for schemes to maximize traffic movement and commercial and industrial growth. But new highways, especially express highways, appeared vital in every plan for urban redevelopment. "Superhighways," argued the head of the Chicago Regional Planning Association in January, 1940, "offer the opportunity to protect the regional value of the central business area" and "to enhance the values of . . . decadent areas and help restore them to a tax paying condition." Parks and playgrounds, as he figured it, would follow. "The city superhighway will not do this work but it will inspire it."[10]

But road building also went ahead according to a second set of standards, promoted by a group of men devoted to the idea that highways should serve traffic, not bolster morale and property values. During the early 1890s, both bicycle club and farm group leaders worked energetically at the local and federal levels to secure road improvements. In 1893, a Good Roads Convention met in Washington, D.C.; by the mid-1890s, both political parties included promises of highway betterment in their state platforms. After 1900, men whose work and hobbies brought them in contact with the automotive and road scenes campaigned for good roads. From 1903 through 1905, Horatio S. Earle, a road equipment salesman and manufacturer and chairman of the Michigan State Highway Committee, traveled the state at his own expense to spread "good-roads fever." In Ohio between 1904 and 1907, members of the Board of Commerce, a businessmen's organization, passed resolutions at annual meetings in favor of highway building. In 1908, they joined leaders of the Ohio State Grange to push legislators to spend more for road construction. After 1908, engineers, dairymen, farmers, and businessmen in Ohio formed a Good Roads Association and directed the struggle. Railroad executives, hoping to increase loadings from areas not served by costly feeder lines, joined state and national road promotors. Between 1902 and about 1913, thousands of American Automobile Association members and executives—men for whom motoring was a social experience—took charge of the national good roads program. By 1914, members of the recently founded American Association of State Highway Officials (AASHO), a network of state and federal road engineers and administrators, headed up road promotion efforts. Getting traffic "out of the mud" served as the symbol around which all could rally.[11]

As much as road enthusiasts talked of speeding up traffic, of schemes for better roads, the legislation they wrote suggests an equally keen interest in centralized management, federal funding, and administration of highway matters by professional engineers. Before about 1900, road building and finance were mostly in the hands of local officials. In Michigan, path masters in each township supervised road work and township officials assessed road taxes. Because county and township leaders in every part of the nation served local transport needs, projects bore little relationship to one another; few worried about interstate coordination. In 1891, New Jersey established the first program of state assistance, agreeing to pay a third of county road costs. By 1915, thirty-nine states had established road departments, themselves usually staffed by civil engineers, and many mandated construction of a trunk line system under their supervision. In 1916, Congress authorized expenditure of $75 million over a five-year period for road construction, provided that states matched on a dollar-for-dollar basis and that state governments administer federal grants through a highway department. By 1917, every state had created a road agency. In 1921, Congress approved $75 million for the next year alone, again to be matched fifty-fifty, and ordered officials to concentrate construction on "such projects as will expedite the completion of an adequate and connected system of highways, interstate in character." Each state, in order to share federal largesse, had to designate 7 percent of its rural mileage for inclusion on the federal aid primary network. By 1925, heads of state road agencies and MacDonald, chief of the Federal Bureau of Public Roads, had worked out arrangements for uniform route markings—those running east–west took an even number, north–south roads an odd one. While as late as the mid-1930s local officials directed road construction, often at federal expense, state and federal engineers increasingly wrote their specifications, supervised their work, and limited their initiatives.[12]

By the late 1930s, state road engineers had developed a sophisticated rationale for maintaining and extending parts of the highway program, for avoiding others, for junking still others. In short, they committed themselves to serving motorists. Because roads were financed from user taxes, or so went this reasoning, they had to produce benefits for them. Motorists' costs, either for vehicular wear and tear or time lost traveling an older route, appeared the best measure of the value of a new road. "Otherwise," warned the head of the American Association of State Highway Officials in January, 1940, "bankruptcy will result."[13]

Efficiency criteria also affected design standards, especially so in the call for expressways and freeways, major roads to which access was limited. As late as 1939, only a few states allowed engineers to block access to roadways. But easy access, as engineers pointed out, encouraged construction of homes and businesses along rights-of-way. Road users financed construction, a Massachusetts engineer told highway contractors in January, 1941, "but the owner[s] of the abutting land, without effort and expense . . ., get the financial benefit and at the same time destroy what others have paid for." A road without access control, he added, soon became "just another city street."[14]

If roads were supposed to serve traffic, as engineers thought, it followed in their minds that state and federal officials could not divert gas taxes for city remodeling, relief, and other non-highway, non-traffic purposes. But in 1936 alone, state officials had shifted 19 percent of gas tax income to schools and such. Diversion, concluded a report of nine engineers presented to a road contractors' convention in 1939, "discriminate[s] against the American motorist." Road building, in their collective judgment, could not be burdened "with any great scheme of slum-clearance, or . . . open to any unmoral scheme of land grabbing." Relief of traffic jams and urban decay was a matter "of improved highway engineering."[15]

Members of the National Highway Users Conference, composed of leaders of trucking, manufacturing, and oil companies, and automobile and farm group executives, joined in celebration of engineering formulas, of engineers' prerogatives. As financiers and as beneficiaries of road construction, they objected to diversion of gas taxes to non-highway activities. So great was diversion in New York, the head of the conference told a gathering of truckers on April 18, 1940, "three or four main trunk highways across the state will not be built." If road users had their way, according to the author of a conference pamphlet published in 1940, "every dollar collected from them in the form of special . . . taxes . . . [would] be credited to them as a highway cost payment." Neither job-making nor urban redevelopment programs fit this framework. "Roads, from the beginning of time," claimed a trucking association official in 1938, "were built for commerce."[16]

During the 1930s, President Herbert Hoover and then President Roosevelt brought still a third set of standards to highway policy making. They focused on the economic impact of highway construction. Massive federal spending for road building and other public works, as Hoover and then Roosevelt envisioned it, would create jobs and thus stimulate the economy. In October, 1930, nearly 7 million were unemployed; by early 1933 nearly 13 million were out of work, roughly 25 percent of the civilian labor force; in 1935, 20 percent, about 10.6 million, remained jobless. On July 21, 1932, President Hoover signed legislation to allow the Reconstruction Finance Corporation to loan $300 million to the states, in part for road building. Between June, 1933, and April, 1934, relief workers employed by the Civil Works Administration under Harry Hopkins constructed and repaired 500,000 miles of roadway. From 1934 through 1937, federal relief officials and Bureau of Public Roads' engineers spent $2.8 billion to construct roads.[17]

But between late 1937 and early 1939, Roosevelt simply could not make up his mind about the relationship between road building and economic recovery. Highway programs "do not provide as much work as other methods of taking care of the unemployed," he told Congress in his budget message dated January 3, 1938, and so federal assistance should "be restored to approximately the predepression figures." For the next federal road authorization, which would run for two years, $125 million yearly appeared about right. On April 14, 1938, as joblessness and the prospect of Democratic losses mounted, Roosevelt urged appropriation of an additional $100 million "only for [highway] projects which can be definitely started this calendar year." Members of the House voted even more funds to build roads, but on May 16, Roosevelt ordered Daniel W. Bell of the

Bureau of the Budget to contact Senators Carl Hayden of Arizona and Kenneth McKellar of Tennessee "to get the total as low as possible." As part of Bell's legislative arm twisting, Roosevelt also wanted him to secure elimination of a 1922 law obligating the federal government for its share of state road spending in advance of a congressional appropriation. This arrangement, in reality, had delegated control of federal highway spending to the chief of the Bureau of Public Roads. "Push for its enactment," he urged Bell. Neither effort, finally, proved successful.[18]

During the first half of 1939, Roosevelt tried to add excess condemnation to federal highway practices. On March 22, he hosted a "no black tie—very informal" stag dinner at the White House for Norman Bel Geddes, designer of GM's Futurama exhibit at the World's Fair. The West Hall was set aside, on Roosevelt's instructions, for a model of Geddes' exhibit, and guests discussed creation of a Federal Land Authority empowered to take extra wide rights-of-way for roads and other public works. Both the president and congressional leaders sought a data and legal base on which to launch their authority. On April 24, Roosevelt told one of his aides to "find out from MacDonald of Highways where . . . I can find anything about . . . buying a wide strip and selling off the surplus land and renting gasoline concessions." By May 15, Senator Hayden, one of those invited to the dinner and a senior member of the Senate Committee on Post Offices and Post Roads, had secured a draft of legislation from the Senate's legislative counsel. Federal road officials, as Hayden had it, could condemn rights-of-way and resell them over a forty-year period to local and state governments. By August, 1940, all that members of the Senate and House roads committees cared to provide was legislation allowing Reconstruction Finance Corporation officials to make loans to purchase rights-of-way as part of the regular federal aid road program, authority which they had in the first place.[19]

Excess condemnation and creation of a Land Authority, for their exotic and esoteric flavoring, appeared in Roosevelt's mind as budget-cutting measures. Unless tolls were collected and land adjacent to roads sold at a profit, he wrote to Budget Director Bell and Interior Secretary Harold L. Ickes on April 1, 1939, "the Treasury is unable to finance even the beginning" of a national superhighway program. Had Congress agreed to relieve MacDonald of authority to approve contracts in advance of an authorization, Roosevelt would have gained a lever on the budget, also at no expense to the treasury. From late 1937 through 1940, what Roosevelt's highway program amounted to was a minor effort to secure votes for Democrats and a major one to enhance his own ability to manipulate the economy.[20]

Beginning on June 21, 1941, Roosevelt and his aides focused exclusively on getting roads prepared for war. Federal Works Administrator John M. Carmody and Lauchlin Currie, an adviser and Federal Reserve Board member, urged Roosevelt to secure legislation mandating priority construction of roads "important to national defense." That day, following their suggestion and using language suggested by Carmody, Roosevelt asked Senator McKellar to secure such an amendment in pending road legislation; and only four days later McKellar, the chairman of the roads committee, wrote Roosevelt that "it gave me pleasure to put this in." On November 24, 1941, as American participation in the world war appeared near, Roosevelt ordered Carmody "to restrict the approval of projects hereafter to those essential to national defense as certified by the appropriate Federal defense agencies." Not again until September 6, 1945, when President Harry S. Truman dropped war-time controls, did normal state and federal road construction get under way.[21]

Between 1900 and 1941, political conflict over taxes and apportionment—basically, who would pay, who benefit—had blocked road construction. But highway and particularly expressway development policy revolved around more than financial squabbles. Planners, engineers, commercial truckers, and the rest, each guided by different training and experiences in their own firms and agencies, simply could not determine in common the direction and impact of American highway programming. President Roosevelt viewed road building in a broad context, assessing highway construction costs in terms of their impact on his own ability to manage economic recovery, but focused on keeping capital outlays low. Urban planners and their business allies—men who operated mostly on the local scene—envisioned road construction as another tool for fostering social cohesion and steady and profitable business dealings. Truckers and road engineers, themselves also local operators, worked in a different, more particular framework. Good highway construction, as they had it, would foster over-the-road transport and serve ever more automobilists, enhancing the profit picture for truckers and the ability of state road engineers to continue to direct highway building. Road construction, for these reasons, was popular with all. But as fast as engineers built roads, and however innovative were their designs, construction never seemed adequate, affordable, or entirely proper to men with such different standards and immense ambitions.

Because social and political differences as well as competition for funds blocked political action, the pace of road building never kept pace with traffic increases nor with visions of faster traffic, social and urban change, and economic improvement. In 1939, according to Bureau Chief MacDonald, "unsatisfactory conditions in respect to sight distance, grade, and curvature . . . [were] in some measure responsible for fatal accidents." Congestion near downtown districts, he continued, "became so bad . . . as to force the abandonment of the route by the through traffic despite its convenient alignment." Around the fringe of central business districts, finally, lay the remains of homes and apartments, now "almost untenable, occupied by the humblest citizens, they . . . form the city's slums—a blight near its very core!"[22]

Roosevelt's freeze on highway building allowed leaders to plan for their own version of construction after the fighting. During the war, the framework in which highway enthusiasts shaped road policy remained about the same. Prewar perceptions of desirable highway impact, of more equitable distribution of funds, still informed political maneuvering. Local business and professional leaders—truckers, planners, and state and county engineers—drew upon particularistic standards to formulate plans aimed at serving local needs. Federal engineers, men who studied road transport and construction with a view toward the national transport scene, entertained their own unique schemes for upgrading traffic and cities. Traditional perceptions, then, judgments based on commercial profitability and on standards developed in day-to-day business and professional routines, shaped wartime planning, new federal road legislation, and visions of postwar America.

2

Planning for Postwar America, 1941-1944

There has been too much confused thinking about highway needs on the part of government officials, legislators, and the general public. The result has been that many state highway departments have not had enough money to construct state highways. This is because unsound practices such as diversion of highway funds to non-highway purposes and dispersion of highway funds to cities and towns have crept into our state governments leaving less and less money for new state highway construction. These are trends that must be stopped; otherwise the future development of our highway system and highway transportation itself will be jeopardized.

Russell E. MacCleery,
National Highway Users Conference,
February 18, 1944

Basic problems—jammed highways and urban decay—continued during the war years. Because of the huge migration of men and women to defense jobs and military bases, because of rationing and restrictions, homes and highways went unrepaired and traffic on roads near defense plants backed up for miles. Other roads fell into disrepair as overloaded trucks and military transport, including tanks, gouged holes. In brief, everything that required substantial outlays of money and material—and what did not—just got worse. While unemployment did end as millions took up posts in the armed services and war factories, many thought that full employment was temporary and leaders continued to discuss the threat of depression-era unemployment lines as if they were a reality.

Even before the United States entered the war, men who studied economics, built roads, planned and managed cities, and operated large businesses began to think about the direction of road transport when peace returned. What, indeed, they asked, were proper and decisive cures for economic and traffic stagnation; how might urban growth be directed along more desirable lines? Construction of new express highways, just as many had claimed since 1900, would create jobs, ease traffic congestion, and help remodel rundown cities.

During the war, those involved in highway construction had several years in which to make plans for postwar building. Initially, authors of postwar plans concentrated on creating a stock of blueprints for highway make-work jobs to offset expected unemployment. But engineers and business and government officials also prepared both massive and fragmentary plans for social, urban, and traffic improvements, each based in part on highways. In 1944, Roads Commissioner MacDonald and his colleagues published a plan for construction of a national expressway network aimed at relieving traffic, creating jobs, and serving as a framework for urban redevelopment. Transportation specialists outlined a national agency staffed by experts who would direct a balanced transport network, one in which rail, water, and air carriers were put on an equal footing with truckers. But farm group leaders and interstate truck operators, proponents of two different points of view, lobbied for their own plans to foster particular and local needs. Farm and rural road enthusiasts—men who celebrated local institutions and special ways of doing things—wanted their own roads attended to. Truck operators, those who were supposed to finance lavish plans, wanted no part of any of them, preferring instead to eliminate federal gas taxes while gaining more federal aid for heavily traveled urban and intercity routes. Truckers and farmers proved most successful in the national legislative arena. In 1944, then, a majority in Congress chose to continue traditional road-building arrangements. They increased aid to construct intercity and farm–market roads, but did add urban highways and MacDonald's Interstate system to the federal aid network. Those who had planned to build new roads as part of a multimodal transport system and others who had hoped to guide urban redesign received neither financial support nor a legal voice in policy formulation.

Between September, 1939, when the war opened in Europe, and late 1941, those planning for postwar road building thought mainly in terms of employment. Highway building, both in the minds of public and private leaders, would create needed jobs. In November, 1939, for instance, President Roosevelt ordered members of his planning group to get ready programs to bolster the postwar economy. In February, 1941, he wanted sketches for roads "put on the shelf, ready to take out when the end comes." Prepare plans, he wrote members of an advisory committee on road building on April 14, 1941, "to utilize productively . . . man power and industrial capacity . . . upon the completion of our defense program." Pyke Johnson, president of the Automotive Safety Foundation, itself a subsidiary of the Automobile Manufacturers Association, agreed. Worthwhile construction projects, he told conventioneers at the 1941 gathering of the American Road Builders' Association, would be required for the "army of men" seeking jobs after the war.[1]

In November, 1941, Congress responded to pressures for postwar road planning by passing the Defense Highway Act. State governments would receive $10 million, provided they matched it, for surveying and planning work on main trunk and intracity routes. Twenty million dollars for planning, Commissioner MacDonald estimated in February, 1943, was sufficient for about $500 million worth of highway building.[2]

By the end of 1941, then, many American leaders had agreed that planning of highway and other projects would prove vital for the postwar years, at least insofar as the employment picture was concerned. Road construction, if done expeditiously and located properly, would open up jobs. Beginning around 1942, a few Americans focused on more grandiose projects. If men were going to be put to work on roads after the war, they should construct a different social, urban, and economic order, not just build highways.

In 1939, President Roosevelt established the National Resources Planning Board (NRPB) within the Executive Office. The chief purpose of board members and their small staff of technicians and consultants was to advise Roosevelt on long-range planning from a national point of view. They prepared reports for him on national resources, demobilization, and economic conditions, especially downward cycles, and served as a clearing house and resource for private and public planners. In general, board members sought to develop plans to raise the standard of living by up to 50 percent, creating an economy of abundance, stability, and cooperation for government, industry, and labor. The postwar plans of Wilfred Owen, an economist and consultant to the board, were a part of this perspective.[3]

By 1942, American highway construction methods appalled Owen. New Deal programs, aimed mostly at creating jobs, had "failed to produce results comparable to the best planning and technological methods." As in any relief program, he argued, morale was low, money was wasted, and construction was more attuned to making jobs than building good, well-located highways. Traditional road construction arrangements, funded locally or through federal grants to state road departments, worked only a little better. Oftentimes, he complained, state grants to local officials "ignored traffic requirements." Sufficient funds were "denied [to] congested metropolitan areas," only to be "lavished upon local rural roads." Some roads, as he perceived it, were "overbuilt . . . beyond any reasonable traffic expectations," and others so poorly funded that "the amount of actual lasting improvements remains negligible." What was needed was a "wholesale revision" of road construction methods.[4]

Beneficiary payment, one of Owen's revisions, promised to reduce road building as well as transport problems. Property owners would have to pay construction costs for minor roads. Through taxes or tolls, truckers and motorists would finance construction of high volume roads. Strict calculation of costs, if Owen was right, would make men cautious. Economic facts, he thought, promised to reduce "political considerations," encourage "a proper utilization of resources," and promote "a more rational selection of alternative transportation methods." Since shipping charges would reflect actual costs, in the long run user fees favored "a more rational selection of alternative transportation methods."[5]

Sound transportation systems were vital for industrial prosperity. They were also devices for shaping communities and for directing regional development. No highway would be constructed, if Owen had his way, unless it was part of an overall land-use and transportation plan for an area. "Transportation planning," he argued, was "second only to 'a basic land-use pattern as a guide in developing the future city.' " But if roads were constructed apart from a plan, if highway engineers were unguided, their work would "dictate what the plan shall be."[6]

Left to themselves after the war, or so Owen claimed, heads of road building and transport groups would continue as before, constructing an inadequate highway system, adding to uncoordinated and costly competition between truckers and railroaders, and furthering an adverse land-use pattern. To bring order and efficiency both to transportation and to urban growth, he recommended creation of a federal land-use and transport agency. In part, agency leaders would see to it that highway plans adhered to a "master transportation plan"; in part, they would coordinate road construction "with housing, agricultural, recreational and other Federal plans." Provided "planning and management" achieved the "high level reached by engineering standards," then transportation, including highway transportation, held out "the promise of unparalleled physical accomplishments."[7]

Members of the planning community, including professional planners and business leaders such as Owen D. Young of General

Electric who were attuned temperamentally to national direction of economic affairs, endorsed the major themes in Owen's program. Most spoke of collecting user charges, but all liked the idea of national transport and highway planning. In order to establish equity between rail, water, air, and road users, concluded authors of a report to accompany Owen's in May, 1942, it might prove necessary for the government to own or lease "all basic transport facilities." Transportation was "a frontier of opportunity," heads of the NRPB wrote President Roosevelt on May 25; managed properly, it would hasten "the unfolding promise of American life."[8]

In other federal and state offices and in private planning agencies, a few spoke of great highways, of quick-flowing traffic, of renewed cities. On April 14, 1941, President Roosevelt had appointed a committee of seven—designated the Interregional Highway Committee—and charged them to plan for the construction of roads following the war. After several years of intermittent deliberations, they produced recommendations for construction of a national express-highway network, one which would serve traffic, provide jobs, and assist with the reconstruction of cities.

Planners, state road engineers, and old-fashioned political appointees made up the committee. Because Senator Lister Hill wanted to boost Bibb Graves' campaign for governor of Alabama, Roosevelt appointed him. The president also selected George D. Kennedy, an engineer and candidate for head of the Michigan Highway Department. According to one of Roosevelt's assistants, Michigan road officials were "politically powerful and control[led] the State government." MacDonald and Carmody picked Rexford G. Tugwell, Frederic A. Delano, Charles H. Purcell, and Harland Bartholomew. Delano was chairman of the NRPB and a long-time member of national planning organizations; Tugwell, former head of the Resettlement Administration, had been responsible for the Greenbelt cities program, a series of federally sponsored new towns built to create jobs and designed to improve living styles; Purcell, a Californian, was State Highway Engineer; Bartholomew headed a large planning firm. Committee members were committed to engineering specifications, traffic flows, and city and regional planning as their form of political expression.[9]

On September 8, 1941, at their second meeting, members of the Interregional Committee outlined standards for route coordinates and financing. Thirty-two thousand miles of limited access roadway, several thousand miles more than MacDonald had projected in 1939, seemed about the right length. This network of roads would service manufacturing and farm areas, obstruct "undue decentralization," and "facilitate the reconstruction of central urban areas." Financing procedures, much as ever, proved more difficult to settle promptly. During the afternoon of September 9, at an informal session of the committee, Bibb Graves suggested earmarking federal automotive taxes for highway purposes. Dedication was uncommon at the federal level, argued others, and was "vigorously opposed by . . . powerful influences within the Federal Government." Why not just call attention to the fact that federal automotive excises were about equal to construction costs, suggested Bartholomew. His idea, according to the minutes of the meeting, "appeared to meet with general approval."[10]

Not until January, 1944, because of the exigencies of war, did MacDonald and members of his Interregional Committee produce a final report. But they still preached a diverse set of goals for postwar America. Traffic service—upgrading the highway system to handle more vehicles at greater speed and safety—came first. Banks, warehouses, hotels, theaters, and government centers comprised the major traffic areas downtown, just as traffic tallies had shown and road engineers had long known, and new roads would have to "penetrate within close proximity." Traffic sources outside downtown—produce markets, stadiums, industrial areas, and the like—were also entitled to "convenient express service." If the new express network served major urban centers and the most important farm counties, as committee members hoped, it would handle about "20 percent of the total of street and highway traffic of the country."[11]

Urban revitalization occupied an equally lofty place in the imagination of committee members. Central urban places, as they pointed out, were "cramped, crowded, and depreciated." Industrialists, shopkeepers, and homeowners had relocated to the suburbs, thus chopping urban tax bases, thus reducing money available for urban services. Construction of Interregional expressways, or so went the reasoning, offered urban leaders a tool to reduce the size of trouble spots and stop others from materializing. Limited access outer belts would discourage outward movement and promote "uniform development of whole areas"; routings nearer downtown were supposed to follow river valleys to eradicate "a long-standing eyesore and blight upon the city's attractiveness and health." If a land-use authority were created and empowered to acquire property for roads, housing, and airports, if its directors could also assemble parcels in "blighted areas," it would promote "a more rational land-use pattern." In brief, Interregional expressways would prove "a powerful force tending to shape the future development of the city."[12]

During the war years, as before, urban businessmen, planners, and politicians blocked out programs for their own postwar renewal programs. In Cincinnati, they were able to build upon work completed by members of a previous generation of planners. As early as 1925, business and political leaders in Cincinnati had prepared plans to guide urban development, focusing mostly on the physical improvement of streets, on garbage disposal, and on accelerating the growth of the city. During World War II, while still enamored of the booster spirit, Cincinnati leaders turned also to broad-scale plans for social revitalization. In 1944, members of the City Council voted to appropriate $100,000 to pay for a review of the master plan by members of their Planning Commission and by two well-known regional planners, Ladislas Segoe and Tracy B. Augur. In turn, they sketched a plan for the reorganization of neighborhoods alongside a remodeled highway-rail network. Shopping centers, post offices, schools, and parks would serve residents of each neighborhood. Two expressways, planned to run through worn-out sections along the river front, would open the way for renewal downtown, including a heliport, marina, modern apartments, and an exposition hall. Such work, as they liked to think, would encourage eventually "the desirable environmental and social conditions" sometimes found in cities of fifty thousand to one hundred thousand.[13]

In most cities, however, men involved in postwar planning lacked resources and enthusiasm for broad-scale revitalization efforts. They tied themselves to local preferences, only laying plans for one or two civic improvements in order to stimulate business. In Los Angeles, the focus was on highways. Beginning in 1942, elected officials and heads of the state chamber of commerce and automobile club in the Los Angeles area met regularly to outline plans for a region-wide expressway network. At a meeting on March 16, 1944, they formed a committee to make ready a report to the legislature, if possible in about ten months. What was needed, the chief engineer of the Automobile Club of Southern California told the others, was "new legislation to provide for a system of freeways in cities and between cities."[14]

Those who thought about postwar highway planning in other cities entertained equally functional proposals. By mid-1943, members of the Post-war Planning Committee of the Toledo Chamber of Commerce had prepared plans for public works pure and simple. The manager of the Sacramento Chamber of Commerce, also in mid-1943, urged preparation of public works plans just to create jobs until industry could retool "to provide permanent employment . . ., stability of property values, and a sound business community." On February 8, 1944, a city engineer in Youngstown, Ohio, wrote MacDonald for information about his agency and "membership fees or other costs." The city government lacked funds to make surveys and develop data for postwar plans, he reported, and so had to rely "on existing organizations for pertinent material."[15] As late as 1944, it seems, civic leaders in Youngstown had prepared nothing.

State road engineers, using federal and matching state funds for postwar highway planning, kept busy during the war years making their own plans. In brief, they outlined route coordinates for expressways. By December, 1943, engineers in the Ohio Department of Highways had blueprinted several expressway networks, including one from Painesville to downtown Cleveland, another from Cincinnati to Lima. About the same time, Kentucky Highway Department engineers had begun to prepare for an express road located in the Covington-Cincinnati area. Michigan road officials, following up on prewar ideas, sketched the John C. Lodge and Crosstown Expressways in Detroit.[16]

Engineers applied the norms of their profession in making postwar road plans. Limited access, as they had argued all along, preserved the investment of motorists; road improvements benefited the users; traffic flow dictated construction priorities; and because of cost and efficiency notions and "local prejudice," the federal government had to finance and condemn rights-of-way. In 1943, as an engineer in the Connecticut Highway Department pointed out, there was "little new" in state plans. After the war, engineers wanted what they had always wanted.[17]

Between 1941 and 1944, a rich outpouring of social thought and criticism and a traditional application of the commercial spirit ran through the minds of men and women making ready for the postwar scene. But visions of the good city and the good life differed greatly. In some quarters, hopes for broad-scale renewal ran high. By building a few miles of expressway and tinkering a little with land use and transportation, many were convinced that they could reshape cities, thus improving the economy, thus curing social problems. The report of the Interregional Committee, the manager of the National Safety Council told road engineers meeting in Chicago on January 29, 1944, should "thrill the imagination . . . of everyone who looks forward to a better America."[18] In other places, however, men planned only to speed up traffic. More highways, especially so expressways, would create opportunities for business enterprise and national and urban restoration of a different sort. That Washington would have to pay, whatever the scale and scope, was about all that expressway, urban, and social uplift enthusiasts shared with one another.

Early in 1943, leaders of the American Association of State Highway Officials submitted a road construction bill to chairmen of the Senate and House roads committees, just as they had done before.[19] Traditionally, AASHO bills served as the basis on which committee members conducted hearings and fashioned highway legislation. No doubt, as participants recognized, whatever emerged from the proceedings—in terms of funding levels, of distribution formulas, of design standards, of control of spending—would favor the plans, sketches, and dreams of one group of men or another.

AASHO leaders, again as always, mostly restated their own ambitions and provided plenty of money to finance them. They wanted a federal outlay of a billion dollars a year for three years, a considerable boost in itself, and federal assumption of three-fourths rather than half of the costs of construction. Apportionment of funds, if members of Congress went along, would rest more on population, less on the size of a state, and the federal government would also pay for construction of separate rural and urban systems. To speed up construction, AASHO executives provided for federal acquisition of rights-of-way, even "prior to approval of title." Costs of parking lots, rights-of-way, and a 40,000-mile expressway network, each a new budget item, also became federal responsibilities.[20]

But the AASHO bill failed to satisfy those with quite different road-building ambitions. Farm road enthusiasts wanted to take care of their own needs first. Beginning around December, 1943, just before hearings on the AASHO bill got under way, rural and small-town officials contacted political leaders to press their point. Heads of the Pratt and Salina, Kansas, Chambers of Commerce, for instance, claimed that the AASHO bill devoted excess funds to urban and interstate road building, not enough to rural local roads. "The urgent need," an official of the Louisiana Highway Department wrote the governor late in January, 1944, "is not for new roads but merely for funds with which to surface and maintain the existing network of farm to market roads." Federal construction standards, moreover, only made matters worse. As of December, 1943, according to Senator Tom Stewart of Tennessee, county road men were "unanimous in their opposition" to MacDonald's administration of local road construction. His "expensive standards," as Stewart related it, were "out of proportion to County needs."[21]

While authors of the AASHO bill had treated road builders in the northeast better than ever, they too complained about management, about funding. William J. Cox, a road engineer and head of the Connecticut Highway Department, led the complainers. Beginning around October, 1943, he contacted road officials in his area plus those in Michigan, Ohio, Illinois, and Virginia, California, and West Virginia. On December 19, 1943, as an illustration of the sort of message conveyed, Cox argued before delegates at a highway conference meeting in Washington, D.C., that the AASHO bill was "hastily drawn without foundation in fact." By seven to three, with five state delegations abstaining, conferees voted to ask for a redraft. Essentially, a majority perceived two problems: first, one of controlling construction in their own states. If state governments paid only 25 percent, Cox wrote a New York official a few days before the conference, "we are 'licked,' from the start." In the second place, apportionment of funds looked unfair. Cox and his cohorts pointed out that urban motorists contributed far more in federal automotive excises than they got back. Under this system, they subsidized "rural sections of the country." By March, 1944, informal negotiations among AASHO leaders proved fruitless, and highway engineers from five northeast states had their own bill drawn and prevailed upon a New Jersey congressman to introduce it. "Discriminations in the [AASHO] bill . . .," according to a telegram Cox and others sent President Roosevelt on April 14, were "unnecessary, unjust, and unbreakable." But if they had their way, the bulk of federal road aid would henceforth be distributed more or less according to a count of motor vehicle registrations. Then, the head of the New Jersey Highway Department informed members of the House Roads Committee on April 24, urbanized states would "be allowed to keep a larger part of their road money until their roads have had a chance to catch up."[22]

Auto and truck manufacturers and for-hire and private truck operators—men for whom gas taxes and road and traffic conditions were serious commercial matters—offered still a third view of the AASHO bill and postwar road legislation. Leaders in the road transportation industry had always argued for more attention to major routes, less to minor ones, and elimination of federal automotive excises, achieving mixed results. Their vision of the postwar road development scene was about the same. In 1943, both to get roads built and to head off federal intervention and unproductive make-work projects for the unemployed, they urged state road department officials to complete plans to acquire rights-of-way. On

February 18, 1944, just as colleagues had done so often in the past, a National Highway User Conference official told road engineers that diversionary and dispersive practices had "crept into . . . state governments," leaving less money available for road building. If cities were to receive increased road aid, "expenditures should be confined to arterial through routes." Whatever program emerged, he added, engineers would have to finance it "on a pay-as-you-go basis . . . at pre-war rates of taxation." Any federal road spending appeared excessive, petroleum industry officials claimed, if state governments had to boost road taxes to meet matching requirements. In July, 1944, after House Road Committee members reduced authorizations in the AASHO bill more or less in line with federal gas tax income, an observer in the trucking industry reported that the efforts of "higher bracket highway user taxpayers" had proved "the most decisive factor."[23]

By the first of August, 1944, proponents of broad renewal plans, of single-minded highway development, and of parsimony in and out of government focused proposals, claims, and demands on Senator Hayden, floor manager of the AASHO bill. He conferred in private with leaders of farm groups, auto clubs, manufacturers, and the U.S. Chamber of Commerce, but only after an executive session of the Committee on Post Offices and Post Roads on August 18, 1944, did he announce a decision. Members of the committee had voted to slice about one-third from the $3 billion AASHO bill, handing over the largest part for more construction on the original federal road network, permitting urban and rural system supporters to split the remainder.[24]

Members of the Senate and House reduced and rearranged further. On September 12 and 15, members of the Senate cut the authorization to $450 million a year for three years, now less than one-half of what AASHO leaders had asked, and returned federal-state sharing to the old fifty-fifty basis. At the insistence of Senator Richard B. Russell of Georgia, they pushed these principles a bit further by dropping federal financing of rights-of-way. Especially in congested urban centers, rights-of-way consumed a considerable part of the road construction budget. Late in November, as part of their own burst of enthusiasm for straightforward road building, members of the House adopted Jennings Randolph's amendment forbidding MacDonald to condemn rights-of-way larger than needed for traffic alone.[25] No aesthetic or urban renewal considerations were going to trouble road builders.

In December, when Senate and House conferees gathered to

adjust differences, few remained. In brief, they authorized $500 million a year for three years and agreed to pay one-third of the cost of acquiring rights-of-way. Population, in this legislation, served as the basis for calculating the distribution of funds to construct the farm–market and the new urban networks, but conferees fixed apportionment of funds for building trunk roads according to the old formula. Provision for designating MacDonald's 40,000-mile national expressway network emerged still intact, but conferees followed the lead of the House Roads Committee and decided to name it the National System of Interstate Highways.[26] They did not fund Interstate construction directly, allowing state road engineers to develop it as they wished from funds authorized for the other networks, no doubt out of respect for strong opinions about the cost and utility of express roads.

On December 20, President Roosevelt signed the Federal Aid Highway Act of 1944. During the war years, truck operators, engineers, and farm leaders had enjoyed the greatest success in federal highway politics. They had managed to keep federal gasoline taxes low and to secure additional aid to construct their own favorite road system. Now there were four federal road networks—the primary or trunk, farm–market, urban, and Interstate—each entitled to help with the purchase of costly rights-of-way, all but the Interstate with their own share of the federal road budget, each looked after by organized constituents. Funding provisions of the 1944 act, moreover, followed traditional federal aid highway practices, since 1921 a set of formulas and arrangements tried and thought true in professional and commercial highway circles. State and federal officials would share expenses, just as before, leaving day-to-day construction to state agencies and private contractors, thus perhaps allowing local interests to secure special audience for particular construction and routing needs. Construction itself, same as ever, remained in the hands of state road engineers, men mostly committed to efficient highway design and quick traffic flow, to serving the areas of traffic generation rather than guiding them.

During the hard days of the 1930s and extending into the war years, Americans in professional and business life had produced thousands of plans for remodeled cities and upgraded and coordinated transportation, many contingent on construction of express highways. If roads were constructed, if all went well otherwise, Americans could expect full employment and national and personal wealth. The 1944 Highway Act, as Senator Hayden had it, was "based on the assumption that the American people cannot enjoy prosperity without an adequate highway transportation system."[27]

But authors of the 1944 act promised fast-moving traffic, jobs and prosperity—no more. They excluded extra wide rights-of-way and parking lots, themselves just bare bones in some of the more gossamer schemes; coordinated transportation and urban development never even had a place in congressional and popular day dreams. So what happened to all those grand plans?

As an interesting counterfactual story, one might argue that President Roosevelt and his advisers had failed to exercise leadership. By refusing to involve themselves in highway politics, they had condemned the Interregional report to library shelves, to study by a later generation of urbanologists and students of aging political fragments. On January 5, 1944, Major General Philip B. Fleming, head of the Federal Works Agency, had urged Roosevelt to promote the findings and recommendations of the Interregional Committee "as promptly as you can." "Retention of initiative in the Congress," Fleming claimed, "depends on that." On January 12, in his letter transmitting the Interregional report to Congress, Roosevelt did urge provision for excess condemnation, but apparently only as part of his old scheme to divert unearned increases in property values to state and federal treasuries. After April, however, following an exchange of correspondence about the apportionment formula for urban areas, available documentation shows no additional executive involvement in highway matters. Indeed, in this version of road politics, presidential leadership was absent.[28] The war took precedence.

But by the war years, men who spoke of professional goals—of road-building procedures and traffic flow standards long thought reasonable by members of the road transport industry—governed public road policy. It was not controlled by presidential arm twisting. Truckers, farm spokesmen, engineers, contractors, and members of the Senate and House road committees insisted that federal road building serve traffic purposes. Although they disagreed about tax and apportionment matters, all could unite to oppose the twin evils of diversion and dispersion. If federal officials would not vacate the gas tax field, as many hoped, it seemed proper for them to invest funds in highway programs for the benefit of quicker traffic flow. Huge transfer payments, moreover, were supposed to promote dynamic economic growth, preventing a return to dull depression days and protecting state road engineers and private contractors from government sponsored make-work. In the face of such ideals, in the face of men organized and committed to agency and business independence and local autonomy in general, planners and their comprehensive plans were no match. By 1944, then, those driven by a national planning impulse, at least in the highway construction field, had been stopped. For the next decade or so, arrangements made in 1944 determined the politics of express highway building, shaping both the problems and direction of over-the-road transportation.

3

The Politics of Highway Finance, 1945-1950

Our work has been delayed this past year by the inability to secure steel and lumber for bridges, machinery, and road building materials, especially cement, which has greatly hampered our postwar highway program. The urban portion of our program has also been delayed because of difficulty in acquiring right-of-way due to the housing shortage.

James R. Law, Chairman,
State Highway Commission of
Wisconsin,
February 3, 1947

Economic growth during the postwar years exceeded wartime hopes dramatically. The years after 1945 were especially prosperous for members of the road transport and highway construction industries. Truck operators increased the size of their fleets and sought new techniques for carrying ever larger loads more rapidly. As truck, auto, and bus sales soared, engineers constructed thousands of miles of roadway.[1]

Yet prosperity did not solve road and traffic problems for members of the highway transport industry, leaders of organized motorists, or shippers; it accelerated them. Before the war, heavy concentrations of traffic along key routes had produced costly and dangerous situations. After 1941, President Roosevelt halted construction, and existing roadway, already in bad shape, was allowed to fall into disrepair. By 1950, after four years of renewed truck, bus, and auto output, traffic and road conditions appeared worse. Human and economic losses were high as drivers jostled one another for space on narrow, winding, and bumpy streets and highways. In New England cities, reported the U.S. Chamber of Commerce in August, 1950, 40 percent of trip time was wasted in traffic jams.[2]

It was axiomatic in trucking and engineering circles that traffic conditions on key routes were outrageous and additional road building was vital. Yet between 1945 and 1950, truckers, farm group leaders, and everyone else with a stake in highway transportation competed with one another for federal road funds. Few wished to finance construction of costly highway systems not advantageous to their own business, professional, and political needs. What seemed more desirable, if truckers or heads of any group could just have their way, was to turn to political leaders with a view toward shifting tax burdens elsewhere and stripping funds earmarked for another network or another government agency to one's favorite road system. Truck operators, for instance, promoted legislation to focus state gasoline tax revenues on highway building rather than schools, welfare, and the like, to eliminate federal gas and excise taxes, and to secure more federal aid to build roads inside their own jammed areas. Farm leaders and their friends in Congress, on the other hand, launched a campaign to collect additional federal aid for farm–market roads. Engineers, for their own part, sold revenue bonds with a view toward boosting short-term income, getting more mileage in place, and thus protecting themselves from toll authorities. During the late 1940s, then, local and unique judgments—about taxes and such—informed the behavior of most who were active in highway politics. Authors of the 1950 Highway Act, as a result, could only recapitulate themes on which road-minded men had agreed in 1944.

During the first year or so after World War II, road engineers, both in Washington and in the state highway departments, focused attention on blocking out coordinates for the Interstate system. Early in 1945, Commissioner MacDonald, anxious to develop a comprehensive highway plan, ordered state officials to submit proposals for their share of Interstate mileage by July 1, 1945. Routes, as he wished, "should be so selected as to form an integrated network." Consultations would follow, claimed MacDonald, and then he would formally designate the system.[3]

Leaders of state road agencies forwarded plans promptly, asking approval for new schemes and financial support and approval to reinvigorate older ones. As early as May 28, 1945, engineers in the Ohio Highway Department had claimed 1,304 Interstate miles, mostly on routes recommended by members of the Interregional Committee, but including fresh items as well. Tallies of vehicle movements on existing roads, including more than 220,000 daily near Cleveland on U.S. 42 and State Route 3, served to justify their application. By August 2, 1947, MacDonald and state road officials had prepared a tentative outline for 37,700 Interstate miles.[4]

But soaring auto and truck traffic soon overwhelmed plans for sufficient road building. Automobile and truck traffic increased at a phenomenal pace. Between 1946 and 1950, Americans replaced older vehicles and added new ones rapidly, forcing up registrations by more than two-thirds. In 1945, about 31 million vehicles of all sorts were registered; in 1946, state officials listed more than 34.3 million; and by 1950, they had registered 49 million, including 8.6 million trucks.[5]

While densely settled states suffered the largest actual increases, rapid jumps really were fairly even around the nation. From 1946 to 1950, New Jersey auto and truck buyers added about 440,000 vehicles. By 1950, Californians alone, among the most enthusiastic motorists, had pushed their total to 4,620,078, an increase of 1.5 million since 1946. Even in Mississippi, where per person income was among the lowest in the nation, auto and truck registrations between 1946 and 1950 increased around 160,000. In December, 1949, Texas highway officials reported that motorists were setting new auto and truck registration records every day.[6]

Inflation, war, and changed auto and truck design, all factors outside highway department control, hobbled efforts to build needed mileage. From 1946 through 1950, state, local, and federal road officials spent $8.4 billion, more than any previous five-year period in American history. But rising prices consumed a good part of the additional outlay. Costs for many construction items zoomed above prewar levels, and went even higher for the unusually expensive parts necessary for urban expressways. Heightened construction standards such as wider radius curves and thicker and wider pavements, all needed to provide safe highways for heavier and faster cars and trucks, added to costs. From the mid-1920s to the early 1950s, traffic increased about 250 percent, but according to Commissioner MacDonald's calculations, service demand on new roads was eight times greater. For a few years after World War II, material shortages delayed construction; during the Korean war, a government order to conserve steel slowed building again.[7]

Rising costs, shortages, and more expensive standards had their greatest impact on the pace of constructing urban and major interurban arteries. Traffic had been slow and unsafe on many of these routes before the war. But because they were the costliest to construct, engineers found it even more difficult to keep them in step with traffic increases. Engineers had to labor seven years and spend

$1 billion to construct 6,500 Interstate miles; 19,000 miles of farm–market roads, built between 1950 and 1951, cost only $232 million. By the early 1950s, so delayed was Interstate construction, engineers speculated that it would require twenty years to finish it in Colorado and Ohio and nearly thirty years in California.[8]

As traffic and costs soared, road engineers and leaders of trucking associations and auto clubs proposed time-honored remedies. All might be right, they thought, provided federal excises on autos, trucks, and fuel were ended and if state governments would spend their gas tax income on construction of key highways. Bond issues might supplement revenues if state officials could not otherwise make do. In short, what truckers and engineers had in mind was more timely road building, security for cherished principles and prerogatives, and lower taxes.

Halting diversion of state gasoline taxes to schools, welfare, and every other non-highway activity appeared the soundest remedy for insufficient road-building funds and traffic congestion. Because of diversion, claimed an executive of the New York State Automobile Dealers Association in 1950, "seventy percent of the state highway system [was] inadequate to meet traffic demands." Rising construction costs, traffic increases, and an aversion to tax increases made it imperative to seal every leak. By late 1949, leaders of organized road users reported strenuous efforts in legislative halls to secure antidiversion legislation. Directors of the Indiana Highway Users Conference, aroused by 1950 at diversion levels and the threat of higher taxes, went further, asking the legislature to return $14 million diverted in the past.[9]

But antidiversion legislation was not enough. Such basic legislation, as National Highway Users Conference Director Arthur C. Butler told a group of engineers meeting at Purdue University in April, 1949, was a "frail reed" in the defense of user taxes. Constitutional amendments, earmarking gas taxes for road work, offered greater protection. By late 1948, twenty-one states had adopted antidiversion amendments, and heads of road user organizations in Maryland, New Mexico, and Wyoming planned to present their legislatures with "Good Roads" amendments shortly. In 1949, directors of the National Highway Users Conference circulated model legislation and copies of amendments to guide petitioners. Vigilance was necessary even after amendments were secured, warned Butler, pointing to an effort in Massachusetts to draw nearly $2 million from automotive taxes to support public transit.[10]

State highway trust systems served as another device to hold on to gasoline taxes. In 1947, for instance, California legislators approved the Collier-Burns Act creating the Highway Users Tax Fund. All gasoline taxes were paid into it, and disbursements were limited to highway construction.[11] Because antidiversion statutes and amendments prohibited expenditure of gas taxes except for highway purposes, such legislation actually created primitive trust arrangements.

Bond financing comprised a second front in the effort to accelerate highway construction while holding down expenses. During the 1920s, state and local officials had sold a great many highway bonds, though their popularity declined rapidly during the depression. Beginning at the end of World War II, as municipal, county, and state politicians sought new sources of revenue to build roads, bonded indebtedness rose swiftly. In 1945, local and state governments issued $47 million worth of highway bonds. In 1946, Atlanta area officials alone sold $40.4 million worth; in 1950, governments issued $521 million worth of bonds for regular road building and repair and another $129 million worth for tollway construction.[12]

Federal road officials were particularly excited about the renewed popularity of bonds. In 1948, according to the head of western operations for the Public Roads Administration (formerly the Bureau of Public Roads), California cities had "reached the legal limit of their bonded indebtedness." But by dedicating gas taxes, at increased rates, to bond repayment, "many urban expressways doubtless could be financed." Bond sales, claimed another PRA engineer, promised "the advantages of a fully improved Interstate system to the American public at an early date." Soon, top officials in the PRA had adopted the bond idea, especially so as "the most attractive alternative" to tollway construction. Toll facilities, as road engineers liked to think, consumed a considerable part of road department budgets and threatened their own authority. Early in November, 1948, Herbert S. Fairbank, next only to MacDonald at the Public Roads Administration, ordered a subordinate to prepare a draft of legislation to allow state officials to use federal road aid funds to pay off bonds. "A principal feature" of the bill, he wrote MacDonald on November 30, 1948, "was to be the provision of means of financing presumably more attractive to the states and financially more sound than toll financing."[13] Bond sales, then, promised a way around debt limits, help to state engineers in battles with toll authorities, and plenty of money and highway building for all.

Road engineers and leaders of highway user groups, just as during the 1930s, also worked diligently to get federal automotive and gasoline taxes reduced. Federal officials collected more from users than they spent on roads and, as the reasoning went, if the federal government dropped automotive taxes, state officials would reimpose them and use the money to build roads. In July, 1947, more than four hundred road users, including for-hire truckers, petroleum distributors, private shippers, and leaders of automobile owners associations, petitioned Congress for repeal of federal gas and vehicle taxes. Road users, they argued, were subject to a "special, class taxation," made unjust because "the burden is determined by the distance the taxpayer must drive to or from his farm or his place of employment." As most petitioners recognized but mentioned less often in public, federal gas taxes were easy to collect, thus federal officials might raise them. By 1950, in any case, road users counted six hundred signatures on another petition.[14]

Ending federal taxes could not mean an end to federal road aid, at least not as Highway User Conference leaders figured it. Linkage of gas taxes to road spending—which they promoted at the state level in the form of antidiversion amendments and the like—was not part of their picture of the federal road finance scene. Indeed, users had discussed "the possible linking of federal automotive taxation to federal highway spending," Conference Director Butler wrote the chairman of the House Ways and Means Committee on July 15, 1947, "and emphatically reject[ed] the theory." What they did agree to, however, "was that the federal government should pay" for the construction of Interstate and major arteries "from sources of general taxation." By 1949, as leaders of trucking and engineering groups wrote newsletters and made speeches at business and professional conferences, all spoke knowingly of recapturing federal taxes, of the federal obligation to continue assisting road construction. "The benefits," a User Conference executive told a gathering of road contractors on November 8, 1949, "to the national defense, the general welfare, delivery of mail and interstate commerce obligate[d] the federal government to contribute to the cost of highways from . . . general taxation."[15]

Since the 1930s, struggles to block diversion and lively campaigns to cut federal automotive taxes and sell bonds were regular features of road politics. Promoters figured, if they were successful everywhere, to reduce congestion and operating costs and outbid ever-threatening toll authorities. Bond income, in the meantime, would substitute for hiked gasoline tax levies on truckers and motorists. No doubt bond sales also postponed some difficult questions about the vitality and even the existence of state road departments in the event of a major federal withdrawal from highway support.

But however shopworn the proposals and whatever their rationale, executives of road user groups and state road departments still endorsed them as standards of good legislation and good finance. On December 14, 1949, directors of the National Highway Users Conference, the major trade association for truck operators and manufacturers, gathered to plan a concentrated assault on federal officials to secure changes in road finance legislation. It was a question, added one participant, of "now or never."[16] But neither woeful petitions for tax relief nor old schemes for financial reform impressed President Truman or rural road enthusiasts.

President Truman and his advisers joined in celebration of conventional assumptions about the virtues of better roads, faster traffic, and highway safety. But they refused to drop the tax on gasoline. In short, Truman and his staff had larger objectives in mind. Between 1946 and 1950, members of the Truman administration subordinated highway building to broad economic considerations. Faced with materiel shortages, especially for home building, and with dramatic price inflation, Truman's primary thrust in highway affairs was to limit construction in the interest of economic management. First, he delayed highway development in order to hold down consumption of scarce materials. Beginning August 6, 1946, Reconversion Director John R. Steelman froze federal highway aid for fifty-six days. According to a tabulation made in August, 1946, by a Highway User Conference official, plans for $500 million worth of highway building would "have to be at least temporarily curtailed." On October 12, 1946, Truman set less stringent limits, ordering Federal Works Administrator Philip B. Fleming to use his "good offices" to get local officials to put off highway construction.[17]

In January and February, 1948, as members of Congress debated renewal of the 1944 Highway Act, Truman's aides sought restraints on federal highway spending in order to reduce inflationary pressures. Because the federal government collected more from road users than it spent on constructing roads, Robinson Newcomb of the Council of Economic Advisers wrote the head of the New Jersey Highway Department on February 2, 1948, "the highway program as a whole . . . [was] deflationary." Only in areas where unemployment was high or roads "deteriorated in such fashion as to cause unnecessary expenditures on tires, car bodies, and gasoline" would Truman's economic advisers approve increased levels of highway spending. At their December 31, 1947, meeting with the president, in fact, they had recommended holding federal spending to $500 million yearly, the figure approved in 1944. To constrain road building further, James E. Webb, director of the budget, convinced chiefs of other federal agencies to recommend a two-year limit on the 1948 Highway Act, not three as in 1944. The purpose of slicing a year from the legislation, as Webb explained it to Truman on January 30, 1948, was to secure "reasonable budgetary control."[18] In the highway field, Truman maintained the Rooseveltian tradition.

Directors of farm groups such as the Grange and Farm Bureau Federation and local, mostly rural truck owners operated inside a different tradition and framework. Accelerated construction of the Interstate system and other high volume roads as well as Truman's financial emphasis both appeared unsatisfactory bases for determining highway needs and distribution formulas. Federal construction standards, which by the late 1940s specified all-weather design, only increased local costs and appeared just as bad. On May 6, 1948, a Farm Bureau executive told road builders and trucking company leaders that access to farms, for production, was most important. The real need was not costly expressways, wrote the executive vice president of the North Carolina Farm Bureau on May 17, 1949, but roads to "get our rural people out of the mud."[19]

Early in 1949, Senators Milton R. Young of North Dakota and John C. Stennis of Mississippi took up the cause of rural road proponents. They held hearings on a bill authorizing $100 million for construction and remodeling of farm highways and creating a separate department in the Public Roads Administration charged with overseeing construction. Only men familiar with rural highway building would staff the department. State and local officials, without advice from federal engineers, would determine geometric standards. No legislation was necessary, Young wrote a Senate colleague on June 10, 1949, provided the bill and hearings prompted MacDonald and his staff to modify policy.[20]

But state road engineers, not privy to the subtleties of Young's promotion, promptly contacted senators in an effort to block the bill. In their correspondence, they spoke of traffic efficiency, of service to motorists, of professional standards. Creation of a rural road division, the head of a state highway department wrote to Stennis on April 20, 1949, would encourage "waste and inefficiency" and endanger "efficient allocation of all funds." Congress had to choose, or so argued the director of AASHO in a letter to Senator George W. Malone of Nevada on May 24, 1949, between "embracing some 600,000 . . . miles carrying about 83 percent of all the traffic . . . or better than 2 million miles of local roads and streets carrying approximately 17 percent." Unless local officials could sustain repair costs during the "economic life" of their roads, then increased federal aid was a "highly questionable policy." Members of the Executive Committee of AASHO, however, worked out an arrangement allowing MacDonald to make a study of local road financing, and Stennis and Young never pushed the bill beyond the hearing stage.[21]

Toward the end of 1949, as the 1948 Highway Act neared expiration, engineers, truckers, and farm group leaders refocused on Washington with a view toward securing their own version of equitable highway finance. On November 21, 1949, at a special meeting held in Chicago, AASHO officers drafted national road legislation and submitted it to members of Congress. Mostly, road engineers opted for more spending on every road network, including federal assumption of three-fourths of Interstate system costs, itself to receive a special $210 million subsidy. State road engineers, if Congress allowed, could transfer one-fourth of their income for primary road building to farm–market construction and use federal aid to repay bonds issued to finance any system.[22] As in bills submitted in the past, leaders of AASHO hoped that members of Congress would spend more to build roads, all while state road engineers directed construction, determined routings, and arranged the details of finance.

Leaders of competing highway groups, men who earned their living by thinking, writing, arguing, and complaining about highway construction and auto and truck transport, liked little about the AASHO bill. On November 8, 1948, a User Conference official had warned of "the peril" to road departments if the federal government picked up 75 percent of Interstate costs. A shift in the sharing of costs, as the head of the Transportation and Communication Department of the U.S. Chamber of Commerce pointed out to the same audience of road contractors, posed a threat to highway department authority and would move the road program toward "the left fork . . . of nationalized highways." Lobbying for highway funds would shift to Washington, where "political pressures" would replace "local needs" as the test of a road program. State road engineers, for their own part, worried less about such matters, one of them writing to Fairbank on July 1, 1950, that he "would jump at the chance of letting the federal government pay this entire bill, or perhaps 90 percent of it, leaving us 10 percent of the responsibility and at least 50 percent of the authority."[23]

Early in 1950, the chairman of the House Public Works Committee, Will M. Whittington of Mississippi, revised the AASHO bill, going a good distance in divergent directions. Interstate system funds would be distributed according to population, a considerable prize for urban motorists, but the authorization itself was sliced substantially. Promotors of strictly rural roads, on the other hand, could look forward to even more money. By April 5, 1950, members of the House Public Works Committee reported a bill that took away a little from Whittington's generous handling of farm–market roads and funded the Interstate at $70 million a year. No one, though, could persuade them to return Interstate costs to a fifty-fifty basis.[24]

On May 19, 1950, although House members had much to say about the Committee bill, they introduced few amendments. J. Harry McGregor of Ohio, a proponent of greater assistance to farm–market roads, asked to strike Interstate funds entirely. From the other direction, Kenneth B. Keating of New York led a small group of representatives from the Northeast in an effort to reduce highway authorizations by 20 percent. This move, he claimed, was made to secure a balanced budget and to stop the "glaring discrimination and tremendous financial drain suffered by New York State taxpayers under these federal-aid programs." By huge margins, both motions failed, and on May 19, members of the House approved the bill.[25]

What was good enough for members of the House was good enough for their Senate counterparts. On July 14, the Public Works Committee, under Dennis Chavez of New Mexico, reported a bill apportioning $70 million for Interstate construction, and offering additional money and promising more influence for farm–market road builders. The bond repayment feature emerged intact, but members of the committee had voted to include cities and counties in the program, for the first time diluting the authority of state road engineers over highway finance.[26]

Beginning August 17, senators from states with the most traffic attempted to alter the committee bill along lines more favorable to their motoring constituents. John C. Lodge of Massachusetts headed a group, also from the Northeast, hoping to put Interstate payments on a population basis. Lodge's formula, or so he argued, would "remove an obvious penalty against those states which most need this federal assistance." Paul H. Douglas of Illinois and Robert C. Hendrickson of New Jersey cosponsored an amendment to slice $125 million from the farm–market road authorization. Senators, voting according to benefits for their own states, turned down the Lodge amendment, 27–58, Douglas-Hendrickson's, 26–54.[27]

Suddenly, on August 17, forty-nine days after he ordered troops to Korea, Truman intervened in the politics of highway finance. Because of "demands for supplies and services in competition with defense needs," he argued in a public letter to Chavez, the Senate bill was "inconsistent with the effort to hold down non-defense spending." It was "essential" to reduce it "at least to the level of $500 million originally recommended in my Budget Message." Federal aid to retire bonds, which encouraged road building, was "particularly undesirable . . . when we are attempting to conserve manpower and materials for our defense effort."[28]

War costs, Chavez told colleagues, justified cuts. But what Truman had in mind for road finance, he reminded them, was what he wanted before Korea. Advocates of economy in and out of Congress, he added, also were pressing for reductions. On August 22, members of the Senate, for whichever reason, voted to lower spending in every category of road aid. Elimination of the special authorization for the Interstate, on the grounds that funding for it was available from other network funds, produced much of the savings.[29]

Now, few differences remained for Senate-House conferees to negotiate about. By the end of August, they had agreed to turn over 45 percent for main trunk building and 30 and 25 percent each for farm–market and urban roads. Interstate construction, such as it was, could draw from those funds. In brief, conferees had struck exactly the arrangement made in 1944. They had maintained "this time-tested ratio," or so ran their report, to "keep faith with the state highway authorities and the users of our federal highway systems."[30]

Between 1946 and 1950, major truck operators, engineers, and proponents of farm road building responded to the traffic mess by modeling their behavior in legislative halls on commercial and professional routines and outlooks. In short, local and highly specialized preferences about, say, tax rates and apportionment of funds set the framework for highway politics. But none enjoyed the strength and skill needed to change opinions or to force opponents to give way. Debates, panels, speeches, accidents, and lost time, income, and lives failed to persuade road enthusiasts and politicians to alter federal highway policy. By terms of the 1950 act, leaders of established groups in highway politics had endorsed the wartime settlement for highway finance. Old formulas, as all understood, could not solve travel, traffic, and shipping problems.

The postwar deadlock over highway finance, even in the face of immense changes in traffic conditions and road needs, only continued conflicts extending back to New Deal days. Everyone defined road transport as vital to prosperity, but all clung to a static picture of highway finance. Farm leaders, interstate and regional truckers, and state road engineers sought revenue for added rural, urban, or expressway mileage, whichever served their commercial and professional needs best, by stripping others. Allowing road engineers to transfer funds from trunk to farm–market building was one phase of this strategy; permissive transfer and bond sales also served to keep gas taxes low.

As highway builders and users strove to protect their pocketbooks, professional standards, and bureaucratic place, they invoked the symbols of other areas. U.S. Chamber of Commerce officials liked federal spending on major highways but disliked assistance to little-used local roads. Aid for farm–market highway building, then, as a chamber spokesman had it on May 17, 1950, was "just plain national socialism."[31]

But postwar highway politics, beginning especially around 1948, included new ideas and different approaches. Civil and traffic engineers updated techniques, trying out complicated apparatus and ingenious formulas for analysis of traffic and land-use variables. Increasingly, they tried to understand road design and construction standards required to cope with modern, high-speed traffic conditions. Beginning in 1951, leaders of highway groups turned to their traffic surveys and updated notions of highway design with a view toward building upon them and restructuring political relationships, writing new road legislation, and getting traffic under way.

4

Project Adequate Roads: Traffic Jams, Business, and Goverment, 1951-1954

Recently we have seen United States Senators and Governors from certain large, wealthy states pleading for the retirement of the federal government from road building and for the repeal of federal gas taxes, on the absurd assumption that all states will impose additional equivalent taxes and be better off on their own. This conclusion is arrived at by trick arithmetic. These people figure what the federal government collects in the wealthiest states, and what it returns. The difference represents profit to these states, and to hell with the rest of the country.

Robert Moses, New York Highway,
Bridge, and Park Department,
December 14, 1953

By late 1951, truckers and highway engineers sensed major changes in the dimensions of the traffic tangle. Congestion, National Highway User Conference Director Butler told heads of the conference on October 11, 1951, had "grow[n] worse." Shippers and carriers complained of food spoilage; employers, of workers arriving late for work; and all, of the inefficiency and tragedy of traffic accidents. According to Butler, the highway situation had "become more costly than we can stand." It was, he thought, a "near crisis."[1]

Political problems—tollway development and competition with railroaders—added a special dimension to discussions of traffic tangles. If road users could not agree among themselves about remedies for congestion, or so Butler believed, railroaders would take advantage of the squabbling to "promote their own selfish objectives." Federal officials would nationalize the highway system, saddling them with higher taxes. Expanding tollway mileage presented another problem. By 1952, more than 600 toll miles were open and another 1,100 or so under construction. A truck operator, Butler reminded conference heads, "likes the road," but "doesn't like the toll." Toll authorities, according to the opinion of many in highway-related industries, would perpetuate themselves, continuing charges and extending their systems.[2]

For years, leaders of commercial truck operators, organized motorists, and state road engineers had lobbied to block tollway schemes, out-distance railroaders, and secure lower taxes. Before 1951, they had concentrated their attention on state and local political bodies, limiting federal level initiatives to biannual pleadings. Often, efforts in Washington and even elsewhere had been ritualistic. But by 1951, so menacing was the prospect of intervention by bureaucrats and by railroaders and so terrible were traffic delays and rising costs and taxes, highway users determined to eliminate obstacles to smooth traffic by remodeling political structures to serve scientific highway development. What was needed, User Conference Director Butler argued, was a program to "get us . . . out of this muddle."[3]

The formal vehicle for this new initiative was Project Adequate Roads (PAR). On May 21, 1951, Butler met with a hundred truckers and truck body and parts manufacturers to discuss road and traffic problems. During the week of October 22, 1951, at a truck operators' convention held in Chicago at the Conrad Hilton Hotel, delegates learned details of PAR. Conference leaders launched it publicly at a meeting held on November 1, 1951, at the Mayflower Hotel in Washington, D.C. In short, they perceived PAR as a national coalition of highway users, manufacturers, public officials, and traffic-research men in the Highway Research Board and Automotive Safety Foundation. While the regular Highway User Conference staff would handle day-to-day operations, the founders of PAR envisioned formation of a series of independent local and state groups. As Butler had explained it to members of his board on October 11, 1951, PAR would become a "national committee for highway improvement."[4]

In most respects, PAR offered highway users another opportunity to express time-honored remedies for road and traffic problems. Construction of additional highway mileage in the most crowded areas, antidiversion legislation, and lower taxes were part of the agenda, just as in the past. One of the few new ideas, a call for a ten-year road construction program, had been passing around building-industry circles for quite some time.[5]

While all these proposals remained important in their own right, systematic rating of highways according to their traffic sufficiency formed the mainspring of the PAR movement. Since the mid-1940s, highway engineers had conducted elaborate surveys of roadway conditions, sometimes broadening them to include analyses of social and economic factors affecting traffic density. By the late 1940s, Arizona road engineers had developed the sufficiency rating idea, inspecting highways by sections and then rating them "on an equal and impartial basis." State engineers, in the PAR version, would study trunk and subsidiary highways and assign numerical scores for structural, traffic, and safety conditions. In the view of proponents, sufficiency rating was an "impartial, unbiased method" of dividing funds between different road networks.[6]

Actually, traditional goals of organized road users were the decisive factors in determining their enthusiasm for the sufficiency system. Subjective evaluations of optimal performance governed scoring. High ratings rested upon the belief that a section of road was perfect for the task assigned to it. A poor road would earn a high score if judged for lighter traffic conditions.[7] Because postwar traffic increases were concentrated on a few urban and intercity routes, they would receive lower scores and preferential treatment. Farm roads, on the other hand, would draw higher ratings and drop in priority for remodeling funds. Sufficiency ratings, in other words, served the political aspirations of road users and engineers, men long anxious to get on with a highway program which stressed congested roads first.

Ultimately, PAR itself was another political movement. Leaders of PAR recognized that revising state and federal road legislation would prove simpler if members of their own industry united around common goals. What they undertook, then, was a hard-pushing, nationwide campaign for concentrated, tax-free, federal road building and more efficient highway programming at the local level. At some point, Butler hoped that PAR would act "as the nation's index finger," directing attention to areas of critical highway need.[8]

By early 1952, PAR had expanded into a large coalition. Its fluid makeup and scientific trappings, especially the sufficiency rating, attracted support from a variety of highway-minded men. At the national level, PAR was more or less an extension of the Highway Users Conference; automobile and truck manufacturers and leaders of trade associations for fleet operators, commercial carriers, and oil companies joined both. Officers of the American Automobile Association represented state and local auto club members in PAR and the Users Conference. Executives of the Chamber of Commerce of the United States were active in founding PAR, though they were not NHUC members. In August, 1952, the chamber's transportation specialist, Henry K. Evans, described PAR as a continuation of the good roads movement in the progressive era. By fall 1952, trucker and auto club leaders in Michigan, Maryland, Illinois, and North Carolina had appointed committees charged with promoting the PAR program before state officials. In December, 1952, the head of the PAR committee in Tennessee reported that the governor-elect had committed himself to the PAR program. At the end of 1952, twenty local PAR groups were active in state highway affairs.[9]

Highway engineers liked the PAR campaign and program, particularly those located in the northeastern states. The chief engineer of Connecticut, Roy E. Jorgensen, joined the Users Conference and PAR, advising members on technical matters and proselytizing on behalf of sufficiency ratings and other common goals. As far back as the 1930s, state engineers and politicians in the Northeast had criticized federal road aid distribution formulas, arguing instead for dropping federal taxes and increasing state excises as a replacement. Because federal officials returned only a small percentage of gas tax revenues collected in their areas, this stance promised solid cash benefits. But highway officials everywhere still expected the Treasury Department to continue to fund construction of important national roads in their states.[10]

Leaders of the Associated General Contractors of America (AGC), a national trade association of road construction contractors, also joined the PAR campaign. In 1948, AGC representatives had begun to discuss with road engineers ways to get federal aid extended across a longer period of time. By 1950, they had agreed to promote a ten-year road-building program because "our highways are more obsolete than they were ten years ago and . . . we are constantly losing ground." Our aim, they concluded at a meeting held in Boston on February 28, 1951, "should be a big volume for each of the next ten years." Members of the AASHO-AGC Joint Cooperative Committee supported the gas tax program pushed by PAR leaders as well, claiming that federal gasoline taxes limited the ability of state officials to raise their own rates. In this matter, road engineers went further than the PAR platform, asking for cancellation of all federal automotive taxes.[11]

Although the PAR coalition was large, members achieved little immediate success. They had not overcome the political strength of President Truman and members of other groups who had always opposed concentrated and tax-free federal highway building. During the early months of 1952, Truman still worried more about inflation than speeding the flow of traffic. Even the arguments of his secretary of the army in favor of greater federal assistance for construction of high-volume Interstate system roads failed to turn Truman's attention from the national economic picture. At the same time, proponents of increased aid for the farm–market network were determined to hold on to their share of federal appropriations. In July, 1952, the head of the Texas Highway Department declared publicly that farm–market systems were more fully developed than rural through-routes and urban roads. The solution to this inequality, he argued, was "to bring the urban and trunk line programs up abreast of, and in balance with, the secondary program." Such a solution was acceptable to many commercial road users; they just did not want to pay for it. By mid-1952, truckers, according to William A. Bresnahan of the American Trucking Associations, already "had a real battle on their hands to avoid becoming the whipping boy for those who are seeking an easy, or selfish or punitive answer" to highway finance problems.[12]

By June, 1952, members of the PAR coalition had not broken the pattern of federal road building. Authors of the 1952 Highway Act largely restated traditional highway finance arrangements, allocating a big share of the federal road dollar to construct farm roads and rural highways of minor traffic importance. PAR leaders even failed to block a small increase in the federal gas tax, though the amount agreed upon—half cent a gallon, boosting it to two cents—was less than feared. On the other hand, this time members of Congress disregarded Truman's plea for a $400 million ceiling on highway spending, and even voted an extra $25 million for construction of the Interstate system alone. But $25 million, at a point when engineers were making estimates in the billions to finish the system, was enough for only a couple of expressway miles in a city such as Boston. On balance, organized road users and engineers had gained little for their efforts.[13]

Directors of PAR remained enthusiastic about prospects for remaking the federal road program. From mid-1952 through mid-1953, in fact, they extended the scope of their activities. Beginning late in 1952, they participated in industry-wide conferences on road problems, coordinated local PAR endeavors, and continued an information dispensing service. Several considerations prompted this flurry of activity. In Bresnahan's view, the industry was locked in a "two-front battle." On the one side, they struggled for acceptance of their scientific approach to road building. On the other, they confronted promoters of misleading notions about trucking affairs. This area was the "propaganda front."[14] Both phases, in the minds of industry leaders, blended into a unified strategy aimed at winning support for scientific highway finance and planning.

Other elements of the Project Adequate Roads movement renewed the struggle as well, repeating familiar plans for reform. State road engineers located in the Northeast still wanted to repeal federal gas taxes. Governors and legislators in quite a few states joined them. They spoke of inherent states' rights and recited the history of federal aid to roads; a few called for abolition of the Bureau of Public Roads (in 1949 changed again from Public Roads Administration) or for reducing it to a fact-gathering agency—all in an effort to hold back from Washington a larger share of the gasoline tax dollar.[15]

By late 1952, the Project Adequate Roads campaign had succeeded in uniting a relatively large number of highway enthusiasts around common themes. But the coalition disintegrated rapidly. Eventually, many truckers refused to rally behind PAR. The fact of the matter was that extensive public relations gimmickry, large conferences, and the mumbo-jumbo of sufficiency ratings had failed to bridge fundamental differences.

Late in 1952, state road engineers broke from the PAR movement, largely in an effort to preserve sources of income and bolster their competitive position against tollway officials. In the past, engineers were eager to get federal taxes dropped. This stance no longer appeared wise. Many had begun to fear that eliminating federal imposts would lead to a disastrous cut in federal road aid without an increase in state taxes as PAR leaders predicted. As a result, engineers in the state highway departments would construct fewer miles while toll authorities built the bulk of expressway miles. "The greatest fallacy" of PAR promoters, the chief engineer of Delaware wrote to the head of the Bureau of Public Roads on April 6, 1953, was "that if all state legislatures re-inact [sic] the two-cent gasoline tax the additional revenue will amount to only $178 million, which is . . . small . . . in comparison to the $32 to $60 billion needed." If the legislators did refuse to reimpose the tax, the new bureau head Francis V. du Pont wrote to the governor on May 25, "Delaware would be in pretty hot water." As before, then, when they lobbied against tying federal gas taxes to highway spending, by mid-1953 most state engineers thought linkage offered financial security and vastly expanded opportunities to build roads. If all went well, they might check toll authorities. By mid-1953, AASHO leaders were urging an annual federal road outlay of $900 million, approximately the amount collected each year in fuel taxes.[16]

In May and June, 1953, as governors, congressmen, and state legislators memorialized Congress and appeared before special hearings on highway problems called by members of the House Public Works Committee, they too endorsed the linkage idea. Some politicians even wanted the federal government to spend its revenues from motor vehicle excises on road construction. In 1953, in response to this growing interest in achieving full linkage, several congressmen introduced bills designed to create a highway trust fund. Federal taxes, both on gasoline and on automobile and truck sales, would be sent to a segregated account to finance federal highway construction programs. Outlays, if they had their way, would amount to about $2.2 billion a year.[17]

Highway engineers and a few politicians were only marginal to the PAR movement, however. Occasionally, they expedited local tasks, especially the engineers with their expertise and reputation as impartial professionals. The real strength of the movement depended on hundreds of auto club leaders and on thousands of truckers and their ubiquitous trade association representatives, men who buttonholed legislators, petitioned congressmen, and appeared before regulatory bodies. But by February 12, 1953, AAA executives at a meeting in Washington, D.C., had drafted a resolution to withdraw from PAR and the National Highway Users Conference. Auto club leaders, in the last analysis, just wanted truck operators to pay more for roads. In Illinois and elsewhere, AAA leaders had managed lengthy and expensive public relations campaigns aimed at shifting a greater part of the road-tax burden to truckers. Only if they paid "their proportionate share," an Illinois AAA official informed club leaders at the meeting, would his own club endorse a highway plan before the state legislature.[18] When AAA officers turned from PAR, in effect they reduced it to a contingent of major truck operators and manufacturers.

But while truckers liked the PAR program in principle, it was not sufficiently inviting to bridge fundamental divisions within their own industry. They split on issues according to the transport needs of the region they served, dividing further between fleet owners, large common carriers, and local, single unit operators. To a considerable extent, smaller truckers opted for state control of road construction and state regulation of transportation matters. In 1935, they had welcomed Interstate Commerce Commission regulation as a way of controlling competition, but wished no additional intrusions into their control of company matters or influence with state officials.[19]

Leaders of the largest trucking industry organizations looked upon federal highway building and industry regulation from still another point of view. Members of this group included Burge N. Seymour of the American Trucking Associations, Roy A. Fruehauf, the head of a truck trailer manufacturing firm, and David Beck of the Teamsters Union. These men and the bulk of truck owners differed significantly in the scope of their respective business operations and their assessment of the regulations best suited to commercial needs. As Beck, Seymour, and Fruehauf wrote to President Dwight D. Eisenhower on January 30, 1953, the industry had matured to the point that it required the "guidance and support" of a single federal agency. This new agency would "dedicate itself . . . to the problem of building a road system in this country adequate to serve the needs of motor transport." Although they did not object to local regulation and state road construction, they had concluded that the multitude of state laws "complicated interestate operations."[20] By mid-1953, then, members of the trucking industry had fragmented themselves, and were unwilling to unite around ambitious promotions calculated to eliminate federal participation in their affairs or to return road construction and finance to state officials alone.

The PAR campaign enjoyed no greater impact on federal officials, many of whom had their own pet notions about highway construction matters. Through most of 1953, as a matter of fact, Eisenhower and his people did not even formulate a highway program. In November, 1953, after nearly ten months in office, Budget Bureau officials awaited a report from Secretary of Commerce Sinclair Weeks on highway funding levels. But the absence of a formal program did not mean that they were not concerned about the relationship between highway building and economic growth and national transportation development. During the second half of 1953, as the economy sagged following the end of the war in Korea, several of Eisenhower's leading officials viewed road construction as a way of creating useful jobs. Highway and other construction projects, Council of Economic Advisers Chairman Arthur F. Burns wrote to Eisenhower on August 11, 1953, would "provide work for those in need of work and help keep the pump primed." During November and December, 1953, by contrast, top officials in the Department of Commerce prepared plans to charge users of federal transport facilities, air, water, and highway included. User charges, they remained convinced, would assure "that each form of transportation will have the opportunity to compete fairly for the movement of the nation's goods." Fuller use of every type of transport, as the reasoning went, would serve national economic development best.[21]

Eisenhower's economists also thought about highway construction, economic growth, and social order. According to calculations made on January 10, 1954, by a staff member of the Council of Economic Advisers, federal and local governments would have to boost highway spending about $3 billion, up to $8 billion annually, "to eliminate the existing backlog." But road construction and public works spending, in this vision, were part of a larger social and economic picture. "Government programs," declared the author of an in-house draft of the Economic Report to Congress dated January 17, 1954, "must be designed to maintain reasonable stability during brief periods of readjustment and to encourage long-term growth." Steady growth, itself "necessary to the . . . survival of America and the free world," was in addition "the best assurance of harmonious social and economic adjustment." Truckers and engineers, however, did not entertain grand notions about patterns of highway building, countercyclical activity, and social stability; their primary interest, recorded regularly since the 1930s, and again in 1954, was in "continuing highway programs."[22]

By the end of 1953, the PAR campaign had failed miserably. Much effort and expense had not shifted high-level federal officials in two administrations to their cause, and truckers remained divided about tax and regulatory matters. Traditional allies such as AAA leadership, highway contractors, and state legislators, governors, and engineers had broken from PAR. Now, old friends demanded that federal officials link gas tax income to road spending, not abolish the tax as PAR leaders recommended. Other than the maxim that more mileage was vital, highway promotors could not agree about much.

Directors of PAR recognized that it was time to change tactics. On October 29 and 30, 1953, at a meeting in Los Angeles, trucking association leaders, the core of PAR, called it to a halt and endorsed the increasingly popular idea of linking federal gas tax receipts to road spending. Faith in the basic fairness and efficacy of beneficiary payment remained strong. But despite the "theoretical merits" of our position, the director of the American Trucking Associations wrote to Congressman McGregor on November 9, 1953, it "was unacceptable as a practical proposition." They too, then, were willing to go along with continued imposition of the federal tax, provided most of the money was tagged "for improvement of the . . . [Interstate] system of highways." But toll finance remained outside reasonable limits. Toll roads relieved congestion, conceded the chairman of the Board of the American Trucking Associations in an address to truckers and Chamber of Commerce officials in Washington, D.C., on December 11, 1953, but they delayed achievement of long-term, more viable solutions to the problems of highway finance. Existing toll mileage stood as "monuments to . . . [their] failure" to find that solution.[23]

Since before World War II, political conflicts over taxes and the relationship of government to social and economic development had stymied elaborate plans and schemes of road engineers and users, economists, and two presidents to update federal road legislation. Between 1944 and 1952, members of Congress, as confused as everyone, had chosen the safest path by incorporating the deadlock. In 1954, each took another turn, quarreling and quibbling for advantage, but a few were increasingly certain that federal road spending should be tied to gas tax revenues.

President Eisenhower fashioned his own approach to highway affairs. In brief, he opted for a sound budget, gentle acceleration of Interstate construction, and a modest increase in his own leverage over economic and road building matters. Federal aid would remain at $575 million a year, what the government was spending already, and he wanted funding authorizations limited to the usual two-year period. "So that maximum progress can be made to overcome present inadequacies in the Interstate Highway System," he told Congress in his State of the Union message on January 7, 1954, "we must continue the federal gasoline tax at two cents per gallon." The money would be used, Eisenhower informed reporters at his February 10 news conference, "to push the good roads program throughout the United States."[24]

Between January and March, 1954, as members of Congress and the administration held hearings and prepared bills, former partners in PAR approached them in disarray, each now anxious to impose his own brand of federal highway building. In mid-February, for an illustration, the former chairman of PAR asked members of the President's Commission on Intergovernmental Relations to recommend $200 million yearly for the Interstate, nothing for the remaining systems. AAA leaders had quite a different formula. On January 29, 1954, one of them wrote to chief presidential aide Sherman Adams that he could accept federal spending on farm–market roads, provided the government took steps to finish the Interstate system "at suitable high standards by 1969." State road engineers could not even agree among themselves. Nine hundred million dollars a year for highway construction, the figure on which they had agreed again at a meeting in Pittsburgh in November, 1953, still appeared acceptable, and most were willing to loosen ties over farm–market construction. But so bitter was feuding, both among engineers and among members of Congress, warned the head of AASHO in letters to his Executive Committee dated March 17, that "I feel that a long continued situation of this kind cannot do other than threaten the solidity of the Association as well as the federal aid road program."[25]

In the climate surrounding road legislation, loaded down as it was with an immense range of finance schemes and emotionally charged issues, members of the House Public Works Committee enjoyed few options. By March 4, 1954, they had determined to find a few more dollars and new construction schedules for all. Proponents of every road system—farm, urban, rural trunk, and Interstate—would get more money, including $180 million for farm–market construction and a whopping $200 million for Interstate. They promised to boost the federal share of Interstate costs to 60 percent and give greater weight to population in figuring allocations to each state, both items obviously calculated to hasten costly expressway building. Because members of the committee were "aware of the fact that the highway users feel that all revenue from the federal . . . tax on gasoline should be expended for highway construction," they made Interstate funding depend "upon the continuation of the present 2-cent . . . tax." On March 8, members of the House took the same tack, approving the committee bill without amendment.[26] Willy-nilly, they had endorsed linkage.

In March, 1954, opponents of one section or another of the House bill scurried about seeking to make things right in Senate legislation. American Automobile Association and Farm Bureau leaders contacted both Eisenhower and members of the Senate with a view toward eliminating linkage, the first on the grounds it "would adversely affect urban highway planning," the latter because "farm use of gasoline has no more connection with highways than the use of fence posts has to highways." The assistant director of the Budget Bureau, however, insisted upon continuing the tax, but claimed it really did not create linkage and was needed in light of "the fact that . . . the federal government cannot afford to expand its activities." What he disliked was the proposed change in the sharing ratio, arguing in a letter, dated March 3, to Senator Edward Martin that in a 60-40 arrangement there "would be less highway construction than is possible under the traditional 50-50 matching ratio." On March 11, 1954, Roger W. Jones, the director of Legislative Reference in the Budget Bureau, ordered White House assistant I. Jack Martin to "make a strong effort to get the 50-50 ratio restored in the Senate and held in conference."[27]

On March 25, members of the Senate Public Works Committee presented their own bill, the thrust of which testified to the continuing deadlock in highway matters. They allocated more money to everyone, though the Interstate was not treated so generously, and then divided it up more or less according to traditional criteria. But as part of a growing interest among government officials in stabilizing the economy through public works spending, they added a section allowing the president to advance by one year the effective date of the bill whenever he determined that unemployment warranted such action.[28]

On April 7, the committee bill encountered few obstacles in the Senate. Attention, such as it was, centered on a recommendation to pay one-half of Interstate funds to each state based on population, just as the House had done. Senator Chavez offered an amendment to return to the usual one-third population and two-thirds land and road mileage formula, a change certain to benefit his own constituents in large but sparsely settled New Mexico. Surprisingly, a few senators from states whose authorizations were lowered under a 50 percent population arrangement, including Hayden of Arizona, voted with the majority.[29]

Administration leaders continued to complain about the road bill, hoping to salvage a cherished principle or practice from the upcoming Senate-House conference. The Senate had voted too much for minor roads, not enough for the Interstate, argued Charles L. Dearing, deputy under secretary for transportation in the Department of Commerce. Another provision, little noticed, allowing the head of the Bureau of Public Roads rather than the secretary of commerce to conduct a survey of highway finance, annoyed him too. "This would tend to defeat the purpose of our office," he claimed; but mostly, it would deprive him of an opportunity to develop his own fairly radical views on toll finance, transport deregulation, and economic development before Congress and the president. Early in April, Dearing contacted Roger Jones at the Budget Bureau. Jones, in turn, directed the matter to Sherman Adams. As a result of these discussions, Jack Martin was ordered to bring Dearing's views to the attention of Senate-House conferees.[30]

At their April 13 meeting, conferees set authorization levels about in the middle. They voted a separate authorization of $175 million for Interstate building and another $700 million for the remaining systems. Federal officials, from then on, would distribute one-half of Interstate funds according to state population and pay 60 percent of Intersate construction costs. The secretary of commerce, as Dearing wished, would study road finance, but conferees decided against allowing the president to speed up spending to create jobs. Debates about linkage no longer mattered, because Congress extended the gas tax in other legislation. On April 14, members of both the houses approved the conference report; on May 6, President Eisenhower signed it.[31]

Members of Project Adequate Roads had hoped to establish a coalition of competing highway users, contractors, and engineers behind a program of scientifically based, tax-free road building. It was a scheme, more than anything else, suited to the needs of cross-country and larger, in-state road users, state road engineers, and auto club officials. But few could agree, within PAR or out, to the details or broader strategy of highway building. The strenuous PAR campaign had not impressed those who had blocked changes for years, and had failed to bridge fundamental differences even among leaders of the coalition. Constituents of farm–market roads and municipal arteries opposed revisions which would return highway construction costs to their budgets. Advancing toll mileage and threatened loss of all federal support frightened road engineers, traditionally loyal to major users on legislation. Road contractors, always happy to construct anything, were less concerned about toll roads, but worried about a loss of dependable financing.[32] High-level federal officials, whether in the Truman or Eisenhower administration, defined their obligation as one of economic stewardship; they bore responsibility, or so they thought, for maintaining economic growth and social cohesion, and highway construction appeared a useful tool. Road users, more particular and local in their tastes, liked government help for the economy, but only so long as their segment was not the wedge of countercyclical efforts. By 1954, what it came down to was that if PAR members wanted greater attention paid to the Interstate system, they had to finance construction of the sprawling federal highway network. Since they were not prepared to spend more than two cents a gallon, there was not much chance of solving the highway crisis.

Postwar urban highway politics revolved around many of the same issues, though the stakes were immeasurably greater. In 1949, President Truman and members of Congress agreed to a national housing and city redevelopment program, one which left key decisions about expressway location and urban design in local hands. In most cities, truck operators and state road engineers persuaded city officials to build expressways with a view toward maximizing traffic flow. But in several cities, urban planners and business leaders predominated, thus succeeding in using their express highways as part of their own plans to revitalize downtown and surrounding neighborhoods. At issue, basically, was the future direction of urban highway building.

5

The Highway and the City, 1945-1955

So the Motorways Plan for the Cincinnati Area is built around the concept of the expressway with its fundamental principle of uninterrupted traffic flow. The expressway has now become indispensable to the general well-being of the community.

Cincinnati Planning Commission, 1950

After World War II, urban businessmen and residents continued to flee to the suburbs, leaving behind declining property values, falling retail sales, and an unsightly collection of decayed buildings and unrented space in the cities. Traffic congestion, since the 1920s a headache for urban leaders, motorists, truckers, and residents alike, composed a particularly critical part of the dilemma. Between 1945 and the mid-1950s, as trucks and autos poured onto narrow streets, traffic tangles grew larger, making the American city an even less desirable place to visit, to play, and to conduct business.

Long before the war, Americans had recognized that urban centers were in miserable condition. Central area businessmen had prepared extensive plans for revitalizing their cities, hoping somehow to bolster sales, real estate values, and morale. Members of the Roosevelt administration such as Rexford G. Tugwell, himself committed both to boosting the economy and improving the human condition, devised even grander strategies for new towns and reorganization of urban social arrangements. Social critics such as Frank Lloyd Wright, Lewis Mumford, Clarence S. Stein, and Benton MacKaye, though operating at the political fringe, prepared utopian plans for regional centers with a view toward restyling social relationships. But whether men proposed broad-scale urban reordering or simply hoped to rescue their investments, each located a new expressway in his plans. During the 1930s, however, social and political conflicts between road users and engineers and planners and members of the housing industry and heads of city governments had deadlocked renewal efforts, leaving library shelves and filing cabinets filled with unworked plans, leaving urban areas to decay further.

Following the war, the shape of urban renewal politics remained much the same. All thought that urban centers ought to be rebuilt and traffic accelerated and that construction of a new expressway or two would aid both processes. But men in positions to act on reform proposals could not agree about which aspect of the urban scene deserved attention first and whose interests should be served. Truck operators and most road and traffic engineers argued for direct route placements to facilitate commercial purposes alone; urban planners and downtown business leaders wanted to use expressways to boost property values and to remodel and reorganize urban centers. As in the past, spokesmen for these conflicting interests deadlocked at the national level. Beginning in 1949, the federal government assumed financial responsibility for urban renewal and continued the expressway program on a modest scale, but administered each one through physically and functionally distinct agencies. But after 1950, in most cities, truckers and state engineers raised traditional symbols—traffic flow, cost, efficiency—and managed to influence those making local route decisions. In turn, decisions about expressway construction helped to reshape American cities.

Members of the urban-planning profession, basically a diverse group of architects, attorneys, retrained engineers, and trained planners, shared several core assumptions about developments in the postwar city. They believed that human personality evolved along more desirable lines within the confines of a neighborhood setting; that sprawl had disrupted urban social arrangements and also lowered property values; and that the central business district, containing the most expensive buildings and most desirable institutions, was most in need of help. Property values could be raised, downtown revitalized, and personalities reshaped, provided only that political and financial tools were available. Express highways, important in the program of every planner, would play a vital part in molding the new order.

Planners, but especially leaders of the profession, had been thinking in these terms for years. During the war, in fact, they had developed an impressive display of plans for the social and physical redevelopment of deteriorating cities. Reconstruction of downtown and neighborhood areas and expensive freeway building were the central proposals. Freeways would speed traffic to and around downtown and also divide neighborhoods from one another. New roads, where they did separate neighborhoods, would act as barriers to residents traveling crosstown but allow easy access to the central business district, thus promoting neighborhood social cohesion and downtown sales.

During the postwar years, many planners continued to think and to prepare plans along these lines. Again they suggested delineation of certain districts for specific purposes, using express roads and rail and mass transit lines to coordinate and tie these subareas to one another and serve as a buffer between factory, commercial, and different kinds of residential districts. At the same time, many planners liked to believe that an expressway system, if they ever got one built, could serve the growth and proper social development of urban regions.[1]

Revival of the central business district, not area-wide renewal schemes, remained the prime concern of urban planners. In 1949, as before, city planner Harland Bartholomew stressed downtown redevelopment. Unplanned growth along the urban fringe, he told students and faculty at Carnegie Institute of Technology in Pittsburgh on May 10, 1949, had brought "economic strangulation" to central city areas. Haphazard growth along the periphery could be controlled, he thought, by imposing city-wide controls on land use. Bartholomew also favored construction of an expanded freeway system for reviving central areas, believing it would bring commercial benefits. Express roads in St. Louis, his own headquarters and one of the cities in which he had undertaken a good deal of work, were aimed outward from the east side of downtown, he informed delegates to a meeting of the American Planning and Civic Association in May, 1949. What he wanted to do was "stabilize values in the business district" and "relieve congestion now too much concentrated on the west." In planning the St. Louis expressway network, moreover, Bartholomew and his colleagues had rejected "existing traffic flow" data in favor of a location strategy which served the economic development of the downtown area.[2]

Not every planner applauded proposals for freeway-building programs. Expressways, in this minority view, accelerated decentralization, contributing to the destruction of the central business district and hastening deterioration of outlying areas. Theodore J. Kent, chairman of the City and Regional Planning Department of the University of California at Berkeley and planning director for San Francisco, claimed that expressways would hasten population dispersal and soon destroy public works' plans. A "costly . . . freeway for the Bay Area," he warned a group of planners and public officials in February, 1949, will be "hopelessly overcrowded and choked by the time it is completed." San Francisco, according to Kent, needed a "region-wide public transit system," one which limited the "impossible demand for more freeways and private automobiles" to a small express-highway network near the central section of the city.[3]

By the late 1940s, then, city planners based their work on several key assumptions. Decentralization was the source of urban disruption and decay. They would have to slow or reverse the process while the remainder of the city, especially the central business district, was rebuilt. Freeway construction, at whatever scale, was to play a role in the redevelopment and recentralizing process.

Occasionally, heads of large manufacturing and commercial firms—men often active in city affairs as part of a commitment to businesslike government and cultural uplift—joined planners and elected officials to promote urban renewal and expressway development. In brief, they believed that the construction of express highways, along with other construction projects, would encourage public and private investments in urban areas. Community-minded men pursued this pattern of attack in Pittsburgh.

Before World War II, Pittsburgh was a dirty, decaying city. Neglect during the early 1940s left it in worse condition; Pittsburgh was a city in decline. Thick smoke, the repeated flooding of polluted rivers, and a virtual absence of new industry and highways distressed many residents. Decentralization and declining property values, the more usual evidence of urban decay, further rendered it an undesirable place in which to live and a less profitable place for business. By 1945, as a consequence, executives of several large companies were planning to move home offices elsewhere.[4]

During the war, business and professional leaders in Pittsburgh determined to arrest these developments. Beginning on May 24, 1943, a coalition of executives at the head of Mellon family business enterprises, led by Richard K. Mellon, professional planners, and leading academicians formed the Allegheny Conference on Community Development. Conference members allied with the political machine of Mayor David Lawrence, drawing in architects, engineers, and planners from the region for technical assistance. Leaders of this group devised a program of smoke abatement, flood control, and physical restoration, all aimed at reviving prosperity in the Golden Triangle, which was the central business district of Pittsburgh and one of the most decayed portions of the city. A projected expressway network and series of parking lots, as coalition members envisioned them, were keys to the success of their program. But their long-range goal was the social and economic revival of Pittsburgh, not faster traffic and a place to park.[5]

Truck operators and truck and automobile manufacturers and professional engineers looked at urban redevelopment plans in different terms. Even before World War II, they had focused on the traffic aspect of urban problems. New highways, in their judgment, were supposed to reduce congestion. "Roads were built for commerce," one of them had argued, and most accepted the idea. After the war, plans for urban revival, broad-scale attacks on transport problems, and anything else which threatened limited revenues or promised higher taxes were unacceptable. Road engineers, as this line of reasoning went, should concentrate tax revenues and talent on untangling stalled traffic.[6]

That traffic flow was more important than urban affairs in general governed the thought and decision making of state road engineers. Since truckers and motorists financed road building, "they should receive first consideration," argued Frank C. Balfour, an engineer and chief right-of-way officer in the California Highway Department. Balfour was aware of urban decentralization, but did not associate these changes with planning in his own department. He focused mainly on relieving congestion by "divorc[ing] through traffic from local traffic." But while improved traffic control and more freeway mileage would speed up movement, "the patient is only half cured if . . . there is no place to park." Ultimately, urbanites needed "sky-scrapers in which to park automobiles as well as sky-scrapers in which to transact business."[7]

Frequently, state highway engineers did speak of road building in the context of an urban complex. Some stressed the importance of highways for retarding decentralization; others claimed new limited-access roads would serve as a framework for downtown redevelopment; still others believed that a good highway network was vital to the competitive position of local business. "A city is a living thing," a New York engineer told an audience of state road officials in September, 1947, and "its vitality depends upon its circulatory system." If roads were "shrunken," inevitably urbanites faced "creeping paralysis"; easy traffic flow, however, would produce "vigor and health."[8]

But in the day-to-day routines of road engineers, as they made decisions about route locations and geometric standards, they focused on budgets and road building, not urban matters. At work and at professional meetings, they talked about traffic, which always got worse, and about new roads to relieve it. By first imposing limits on their own professional roles, and no doubt agency heads and politicians imposed bureaucratic and political limits as well, engineers channeled themselves into a narrow road and traffic orientation. According to notes of a discussion among several engineers held at a professional meeting at the University of California at Berkeley on February 2, 1949, all judged that "highway departments are committed to the problem of providing facilities for the moving of existing traffic, but that influencing land-area development . . . is outside the scope of a highway department."[9]

Between 1945 and 1949, top-level officials at the Public Roads Administration (PRA) promoted plans both for urban redesign on a grand scale and for massive urban expressway systems designed to hasten traffic. Indeed, this task was not an easy one politically, since highway engineers composed their principal constituency. But Commissioner of Public Roads MacDonald and his colleagues, were their proposals accepted, hoped to establish themselves as directors of urban resuscitation and American road building.

Their initiative developed slowly over time, mostly in response to opportunities in postwar politics. Early in 1945, MacDonald had asked state highway officials to submit plans to him for construction of the Interstate system. Consider the usual criteria—expense, maintenance, and motor vehicle usage—they were told, but also pay attention to "economic and social values." Patterns of land use, if a conflict developed, should receive greater weight in their route calculations "than the existing numerical volume" of traffic.[10]

What PRA officials recommended was a design for urban expressway systems aimed at promoting social as well as traffic advantages. First, engineers would place a belt road around downtown, with adequate off-street parking nearby. Next, they would construct a series of arterial highways running to the outer edge of the city, itself encircled by an outer-belt highway. The system, when finished, would appear from an aerial view to form a hub, spokes, and wheel. The inner belt would guide traffic around the edges of downtown, reducing congestion in the central business district, allowing motorists who entered the core to channel themselves closest to their destinations. The arterial routes, though an additional asset to traffic movement, would also link disparate sections of the city. An urban expressway network, if well conceived, Assistant PRA Commissioner Fairbank argued before a meeting of engineers in Milwaukee on April 18, 1945, "will serve and can promote a generally beneficial urban development"; if laid out poorly, it would "thwart . . . desirable city growth and change."[11]

Soon, PRA engineers were busy promoting their version of good expressway and urban renewal planning. "Blight had already attacked many cities," western operations chief Lawrence I. Hewes told a meeting of business and political leaders at the Commonwealth Club of San Francisco on December 7, 1945, and "continuation of trends . . . may lead to insolvency of the city." But although automobile travel and road building had accelerated decay and decentralization, "the modern expressway can help preserve the city" by "allowing convenient access now found only in suburban centers." It was important to coordinate urban renewal and expressway building, Hewes added in an address to club members on August 5, 1946. Both, he thought, should become part of a slum clearance program getting under way in California. Because express roads would penetrate areas with "obsolete buildings and lowered property values," engineers and planners would have "to work hand in hand to obtain the maximum benefit in cleaning up those blighted sections."[12]

A shortage of housing after the war forced MacDonald to restrict his plans. The postwar housing scene, according to one observer, was a "national calamity." In order to conserve resources for home construction, then, President Truman limited federal highway spending and imposed limits on the use of materials for road construction.[13]

The housing shortage, as things worked out, encouraged MacDonald and Federal Works Agency head Philip B. Fleming to seek a legislative mandate to go ahead with their expressway and urban redevelopment notions. Several bills were pending before Congress to correct the housing problem. As early as November, 1947, Fleming and MacDonald perceived upcoming legislation as an opportunity to insinuate themselves into the urban renewal field, retain control of highway building, and direct both toward a broader program of urban redesign.[14]

On December 21, 1948, Fleming wrote to Truman that contemplated legislation allowed only a narrow program of slum removal and replatting. Such a limited approach, in his opinion, would "defeat the basic purposes of the program." What was needed was a mandate for a coordinated approach, all centered around him in the Federal Works Agency. As he pictured it, MacDonald could continue his urban road-building program, thus quickly eliminating "thousands of substandard houses." Leaders of the Public Buildings Administration, another of Fleming's subagencies, would plan federal buildings as part of the civic center redevelopments taking place in many cities. Executives in the Bureau of Community Facilities, still another Federal Works subagency, would loan funds for planning additional public buildings in redeveloped sections. New express highways, in the final picture, "will be the framework of the redeveloped city."[15]

By April 20, 1949, Truman and his aides had reviewed Fleming's proposal. While the housing bill, he wrote Fleming that day, did "not cover as much ground as may be desirable," it seemed wisest to "hold to the provisions . . . for the present." For several years, housing bills had encountered trouble in Congress, and Truman decided to press ahead to assure passage. Current legislation, he reminded Fleming, represented an "advance over anything which we have been able to do in the past."[16]

Truman's decision to opt for an urban housing program shorn of freeways contributed to the broad division of opinion and interest over the direction of urban affairs. As of midcentury, proponents of two approaches to urban and traffic problems contended for attention and support. City planners and downtown businessmen looked to expressway construction and maybe upgrading of mass transit as techniques for both reviving their cities and speeding traffic. In planning transport improvements, Harland Bartholomew and Richard K. Mellon downgraded traffic flow studies, instead plotting route coordinates according to expansive visions of physical renovation, fast-moving traffic, and downtown business revival. Downtown business leaders and planners, in optimistic moments, hoped that freeways would help to create more compact urban subcommunities. These island neighborhoods, in turn, would serve as focal points for a more harmonious social order. The second approach, promoted directly by engineers and inadvertently by organized road users, emphasized a simple highway solution to traffic problems. Frank Balfour of the California Highway Department and Arthur Butler of the Highway Users Conference rejected proposals to tap gas taxes for other purposes or to set highways outside major traffic corridors. In their day-to-day business and professional work, they concentrated on the advantages for vehicular movement of more road mileage, especially high-speed expressways. Fleming and MacDonald tried to combine elements of each approach, leaving themselves in command of federal urban programs, but President Truman blocked them.

Both Presidents Truman and Eisenhower and members of Congress decided to recognize these competing visions, claims, and needs in separate programs. The city planning and urban housing impulses matured into the Housing Acts of 1949 and 1954. Between 1944 and

1951, the expressway program received irregular funding as part of the regular federal road aid program. In 1952 and 1954, road users and federal officials agreed to finance construction of the Interstate system from special authorizations.

But federal funding failed to affect political battles between local competitors. The Housing Acts did not obligate city officials to proceed along any particular renewal path; direction of highway routings and construction remained in the hands of state road engineers. Beginning in 1950, then, the federal government financed the program of either city-wide planners or highway enthusiasts. In short, it was a matter of which faction exercised greatest leverage on the local scene. This flexibility meant that local political arrangements influenced the location of urban expressways, thus allowing engineers, truckers, or planners to remodel American cities.[17]

In New Haven, Connecticut, proponents of city-wide planning managed to predominate. After World War II, by the usual standards, New Haven was a disintegrating city. Professional men and businessmen moved to the suburbs, or New York, and retail operators relocated from downtown to outlying shopping centers. In 1952, Sears, Roebuck vacated its New Haven store for a suburban site and Gamble Desmond, another department store, closed its doors. To maintain tax revenues, city officials overvalued downtown property, thus accelerating the destruction of economically marginal buildings. Tangled traffic and the Oak Street slum, a conglomeration of uncollected trash, poor housing, and rats made downtown New Haven an unprofitable place for business and an unpleasant one for fun-seekers. Post–World War II New Haven, like many American cities, was in a state of "total crisis."[18]

A coalition of concerned business, professional, and political leaders directed revival along comprehensive lines. Beginning in January, 1954, Mayor Richard C. Lee was the dominant force, but planners from Yale University and local executives joined in actively. Their early efforts for urban growth and redevelopment rested on a plan prepared in 1941 by Maurice Rotival, a planner on the Yale faculty. That eradication of urban decay depended on control of traffic was Rotival's overarching theme, itself an old idea by the late 1930s. On the basis of this scheme, updated in 1951 to take Interstate routes into account, Rotival and Mayor Lee's plan director, Norris Andrews, selected two downtown districts and six residential areas for renewal. Soon, they had marked out about one-fourth of the city for demolition and reconstruction.[19]

New express highways, once renewal was finished, were scheduled to serve varying needs according to the character of neighborhoods through which they passed. In residential neighborhoods, they would pose a buffer against adjacent manufacturing and commercial districts. Freeways, in the view of Mayor Lee and renewal enthusiasts, would also encourage travel downtown and make movement between industrial zones easier.[20]

But if civic leaders in New Haven promoted expressways as buffers, frames, or the like, what these men sought most of all was broad-scale physical restoration of their city and recreation of viable neighborhoods. Often, they located new roads to present maximum advantage to renewal plans, leading occasionally to intense political conflict between Mayor Lee and engineers in the state road department. In brief, Mayor Lee and his cohorts perceived expressways, along with usual street widening and traffic control programs, as tools for reshaping physical and social environments.[21]

In St. Paul, Minnesota, on the other hand, political leaders chose to construct their express highway system according to traffic patterns. The usual run of urban problems burdened businessmen and motorists in St. Paul. Between 1945 and 1955, retail sales downtown dropped 15 percent and traffic congestion grew worse. Downtown store owners and professional men, as in other cities, sought new locations in the suburbs.[22]

By the late 1940s, truckers, engineers, local politicians, and businessmen agreed that a long-proposed expressway was vital to St. Paul. The route chosen, most acknowledged, should circle downtown, pass near an industrial section to the west, and then continue to the University of Minnesota, all areas of considerable traffic. In 1949, engineers in the Minnesota Highway Department conducted traffic flow surveys. Data, they argued, showed that a route located slightly to the south of major traffic areas would allow the quickest trip downtown and cost least to construct in the long run. Motorists on nearby streets, they added, would use the freeway, thus reducing local congestion. Cost and traffic factors, then, justified their choice.[23]

George Herrold, city planner of St. Paul, wanted to serve different interests. Twenty-two railroad lines plus boulevards sliced the city into tiny settlements incapable of maturing into desirable communities, he claimed, and coordinates laid out by state engineers promised only more cutting. Herrold himself plotted an express road to run north of downtown, the industrial area, and the university. If his corridor did not service existing traffic patterns, Herrold believed that was its virtue. By constructing an expressway outside regular traffic patterns, he figured to reduce congestion and increase utilization of undeveloped land on the northern part of the city.[24]

City officials opted for the highway engineers' southern route, weighing several factors in their decision. Herrold, a man in his mid-eighties, had cultivated few friends at City Hall and lacked a staff to collect sophisticated traffic data. Truckers and merchants based in the industrial section favored the closer, southern route; commuters who resided at the western edge of the city preferred the time-saving promised by the engineers' location; and downtown merchants just liked the promise of more business. Financial considerations weighed heavily in their decision too. Herrold's northern route, road engineers warned, would not attract enough traffic to merit their approval and their money. If officials selected his location, they would have to finance construction and maintenance from local taxes. For cost and traffic reasons, leaders of St. Paul chose to base their renewal program around an express highway.[25]

The dominant themes in postwar urban politics, endorsed by politicians, truckers, engineers, and planners, were physical redevelopment and economic growth. Best, all seemed to think, to eliminate decay and hasten development by constructing express highways. A majority of urbanites joined in celebration of highway building. During the 1940s and 1950s, a time when little money was available from Washington and state houses for costly expressways, urbanites approved gigantic bond issues to get the roads they wanted. In 1947, for instance, Kansas Citians voted for a $12 million highway package; in 1955, urban governments sold $310 million worth of bonds for highway construction.[26]

While expressway building was popular, indeed, neither planners and business executives nor state engineers could join together to support one route location or another. It was possible and desirable, claimed planners and business allies, to direct the course of growth by channeling expressways between neighborhoods and building urban renewal complexes nearby. Upgraded neighborhood settings, remodeled as in New Haven in conjunction with the road program, promised still more prosperity. Faster traffic, new neighborhoods, and vast investments downtown, should things go as planners hoped, promised an efficient city and a wealthier and a more harmonious social order.

Truckers and state road engineers approached urban renewal and highway construction from a narrower frame of reference. They preferred efficient traffic movements to an efficient city; and they thought that highway builders should serve existing investments and contemporary economic and social institutions, not redirect them.

What effects expressway construction might have on urban growth and social relationships played no part in their calculations. The ultimate source of division, then, was between engineers and others who favored unlimited economic development and planners and their friends, men who idealized a process of directed, systematic growth and an informal program of social control.

Because President Truman refused to promote a broad-scale attack on urban problems, because Congress funded both highway building and urban renewal independently, relationships established on the local urban scene determined the direction of renewal in each city. In New Haven and Pittsburgh, key members of the business communities perceived problems on a fairly broad scale and identified the prosperity of their firms with the economic health of the entire city. They joined political leaders to mobilize resources such as planning staffs, themselves composed of men who identified the implementation of professional standards with the welfare of their city or even their region. In addition, there were powerful figures —Mayor Lee, Richard K. Mellon, Mayor Lawrence—who gave overall direction to recovery. They coordinated local highway and renewal programs and employed the prestige and influence of their posts to attract funds and to secure cooperation from competing business and political leaders.

This combination of public and private leadership, of men anxious to work for city-wide renewal, simply did not exist in most cities. In St. Paul and in cities such as New York, neither businessmen nor political leaders were comfortable with broad definitions of prosperity. Merchants as well as motorists and truckers considered highway building in terms of personal profit and convenience, perceiving no need to project themselves beyond those considerations. The political structure of St. Paul—a "weak" mayor system—made it difficult for an incumbent, however articulate, resourceful, and well intentioned, to impose an urban plan. By their non-decisions and decisions, leaders in St. Paul, as elsewhere, allowed road engineers to test Wilfred Owen's aphorism. If engineers were permitted to build without reference to an urban plan, he predicted in 1942, then their work would "dictate what the plan shall be."

But American highway building, for whatever purpose, had been stunted at the national level for years. Major road users had launched strenuous political campaigns, hoping to dislodge farm–market road building from the federal payroll while securing tax-free, expressway construction for themselves. Local road advocates, for their part, wished to charge interstate truckers and motorists with an even greater share of farm highway expenses. Larger questions concerning the pace and direction of economic development, local autonomy, political independence, and professional judgment had permeated these debates and fractured highway politics further. Beginning in 1954, President Eisenhower took his turn at trying to loosen traffic and to impose order on the economic and social system.

6

Dwight D. Eisenhower and Express Highway Politics, 1954-1955

America lives on wheels, and we have to provide the highways to keep America living on wheels and keep the kind and form of life that we want.

George M. Humphrey,
Secretary of the Treasury,
May 2, 1955

For years leaders of the highway transport and road construction industries had argued and complained about the pace, direction, and financing of new highways, particularly costly express highways. Farm group executives, heads of in-state and national road-user associations, and state and federal as well as county engineers had pressed one another to pay more attention to their favorite road system. All had worked aggressively, moreover, to shift the financial burden of highway construction from themselves to general revenues, to taxes on other road users, and to property owners.

But if so many had been busy trying to alter legislative and road-building arrangements, ultimately none had proved very successful. As truckers increased fleets and Americans everywhere purchased more cars, it was increasingly difficult for road officials—themselves always short of funds—to construct enough mileage in key spots. Not until sufficient revenue was channeled into construction of roads located in packed corridors would traffic flow smoothly, or so trucking and motorist association leaders liked to argue. Yet since the 1930s, men anxious for roads had deadlocked over highway finance, administration, and apportionment, thus freezing federal highway formulas. Early in 1954, in fact, leaders of competing groups and President Eisenhower had endorsed added attention for the high-volume Interstate system only when truckers agreed to subvent the entire federal aid program, including in particular greater sums for little used farm–market roads.

By spring 1954, President Eisenhower and his advisers were not satisfied with these arrangements. He recognized, as they did as well, that the existing level of highway construction failed to solve the traffic crisis and failed to serve as a long-range foundation for economic growth. Eisenhower took charge of government road planning, but this time he asked Lucius D. Clay to coordinate it. Clay accepted responsibility for writing road legislation acceptable both to the president and to a majority of organized road users and members of Congress. Clay's bill never gained sufficient votes, mostly because congressmen and road enthusiasts perceived highway finance in local terms, not as part of their own contribution to national economic planning. In the end, the president could not disrupt the pattern of coalition building which had dominated highway politics since the era of Franklin Roosevelt.

Between January, 1953, and early 1954, Eisenhower's top-level officials had failed to agree about the essentials of a remodeled highway program, one actually sufficient for handling the traffic upsurge and bolstering the economy. That more road building would halt the post–Korean War downturn in the economy seemed reasonable enough to most, but for the specifics of a program all awaited the report of a presidential commission and lobbied for pet schemes. For instance, Robert B. Murray, Jr., the under secretary of commerce for transportation, believed that state-financed toll roads could meet traffic needs. On the other hand, Francis V. du Pont, the new commissioner of the Bureau of Public Roads, argued for 100 percent federal financing of the Interstate system. Bureau officials—traditionally cosmopolitan and optimistic men—were also promoting establishment of an office of under secretary of commerce for highways.[1] They sought to direct development of the national highway system, and a subcabinet level ranking seemed a propitious spot from which to nurture such ambitions.

At an April 12, 1954, meeting in the White House, President Eisenhower reorganized government road planning and tried to impose his own views on federal highway programming. Since at least mid-February, 1954, Eisenhower had believed that the federal government should boost road spending in order to accommodate traffic. More automobiles, he thought, meant "greater convenience . . ., greater happiness, and greater standards of living." Now, he wanted Sherman Adams, his chief assistant, and Arthur Burns, head of the Council of Economic Advisers, to coordinate a search among government officials for methods to accelerate the federal highway building program. Eisenhower himself was seeking a " 'dramatic' plan to get 50 billion dollars worth of self-liquidating highways under construction." In terms of construction priorities, he thought the federal government ought to devote greater attention to the Interstate system, to roads from airports into downtown areas, and to access roads near defense installations. While he would condone federal loan guarantees, an expanded road program could not be allowed to upset the federal budget.[2]

Soon, several top officials were busy developing plans. But each chose to interpret Eisenhower's instructions differently. Old fissures between administrators reappeared quickly, then, as each produced different proposals to finance and control construction and apportion funds.

Economists and those interested in economic planning evaluated road-building formulas in light of long-range business trends. Traffic relief, they thought, was necessary to encourage growth. Sufficient roads for traffic, predicted one economist, would "mean the difference between a prosperous enjoyable economy, and a more restricted, harassed one." At the same time, massive road construction, if timed properly, offered a useful device in a program for controlling economic swings.[3]

The head of the Public Works Planning Unit in the Council of Economic Advisers, General John H. Bragdon (U.S. Army, Ret.), drew up plans for a more centralized program. He hoped to secure Burns' support for a policy of firm federal direction of highway building. The secretaries of commerce, defense, and the treasury, as he envisioned it, would sit as a board of directors of a National Highway Authority, assuming responsibility for federal road construction and finance. Highway construction was a national obligation, he believed, not something to be divided or left to local governments.[4]

Even in this rudimentary stage, Bragdon's proposal stood in contrast to the workings of the traditional federal highway program. It had emphasized the supervision by engineers in the Bureau of Public Roads of locally initiated projects and roughly a 50-50, federal-state sharing of expenses. At best, moreover, Bragdon's plan downgraded state and local highway officials to administrative agents of the National Highway Authority.

Sherman Adams took the lead promoting a second approach, one more consonant with American road-building traditions. He too figured that executives of a national road authority would finance construction, allowing a small subsidy for toll roads not fully solvent.[5] But Adams entertained few plans for revising federal road-building arrangements.

In April, Adams turned to Bertram D. Tallamy and Robert Moses, leading New York road officials, for details of an acceptable plan. Around May 1, they turned in a report which adhered to the outlines of Adams' views, created a device to raise funds for their own use, and insured local and state authority in the highway construction field. The secretary of the treasury would head a Continental Highway Finance Corporation with the secretaries of defense and commerce serving alongside him as a board of directors. They would look after financial matters. Daily operations would remain under the direction of bureau and state road engineers.[6]

By May 24, Bragdon had prepared a critique for Burns of the Moses-Tallamy Plan. Use of bureau officials, and especially the bureau head as operating chief, would constrain efforts; he also objected to continued involvement of local officials in construction. Even a request by Moses and Tallamy for a moratorium on highway legislation until early 1955, pending a report on traffic conditions and finance, appeared undesirable; it would consist of just "more words." Throughout May and June, as men in other government offices gathered to discuss road and traffic problems, the same sort of conflicts took place.[7]

Adams and Burns recognized the deadlock and had begun to seek techniques to work around it. As early as May 11, Adams thought it would require some "power plays" before Meyer Kestnbaum and members of his Committee on Intergovernmental Relations would approve the Moses-Tallamy plan. Kestnbaum's approach disturbed Burns too. Originally, Eisenhower had relied upon Kestnbaum and his committee to prepare a federal highway program. After reading Kestnbaum's views in a memo from Bragdon, Burns inquired: "Where do we go next?"[8]

In July, Eisenhower halted this internecine debate, directing his aides to begin another search for a solution to the traffic and political tie-up. On July 12, in a speech to the Governors' Conference (delivered on his behalf by Vice President Richard M. Nixon), Eisenhower advertised his interest in stepping up the rate of highway construction. Governors and heads of interested groups, he told them, were invited to participate in planning.[9]

On July 22, Burns advised Eisenhower to create two committees charged with responsibility for coordinating federal highway planning. Burns urged him to create a federal Interagency Committee and a five to nine member extra-governmental group designated the President's Advisory Committee on a National Highway Program. Members of the Interagency Committee would consider economic requirements for a national road program and then submit a construction and finance plan to the president's advisory committee. State highway officials, road users, and governors would present their own proposals to the advisory committtee as well. Members of the advisory committee, as Burns projected it, would sort out these plans and prepare a report for the president's use in his 1955 State of the Union Message. Next, they would launch "an aggressive campaign" for congressional approval of their own and Eisenhower's plan.[10]

Eisenhower assembled two road study groups, each along lines suggested by Burns. On August 20, the president ordered the secretaries of the defense, treasury, and commerce departments as well as the director of the Budget Bureau and chairman of the Council of Economic Advisers to designate representatives to an Interagency Committee. Eisenhower assigned Dr. Gabriel Hauge, an economist and one of his assistants, to serve as liaison. Bureau of Public Roads Commissioner du Pont would take the chair. To head up his advisory committee, Eisenhower called upon General Lucius D. Clay, a friend since days in the European theater and by 1954 the president of Continental Can. Clay would collate the views of state highway officials, governors, road users, and members of the Interagency Committee into a coherent proposal.[11]

To Adams and Clay, Eisenhower delegated responsibility for selecting additional members of the advisory committee. They ignored recommendations from colleagues and picked men with experience in related industrial groups. But the men chosen also appeared capable of taking a national view of highway development, not just a regional or industry-wide perspective. According to Clay, consultants such as Robert Moses, an AASHO representative, and a few others would represent "special interests."[12] By late August, then, another effort to accelerate and restructure the national highway program was under way.

State highway engineers, governors, and truck operators reacted enthusiastically to the president's offer to submit plans and to the prospect of more highways. On August 23, Robinson Newcomb, a former member of the Council of Economic Advisers, reported "a meeting of minds" between some AASHO members and the governors. Leaders of auto and truck associations, men who in the past had tried to overturn federal road policies, also were reviewing positions. In fact, Newcomb perceived "a ferment" which had not been present "since the 20's."[13]

But within government circles, old conflicts resumed. Points of debate were roughly what they had been prior to Eisenhower's reorganization of the search committees. Indeed, it was as if the president had never intervened.

A few months earlier, Commissioner du Pont had gone to several state highway officials, including Tallamy of New York, for outlines of a national highway authority. Around the end of July, they had turned in a report much like the one prepared for Adams several months earlier. Authors of both recommended creation of a road authority and urged financing of Interstate construction by dedicating federal gas tax income to repayment of a bond issue. But the thrust of their plan was to tighten relationships between federal and state road engineers and augment everyone's revenue considerably.[14]

On September 9, at the first meeting of the Interagency Committee, du Pont submitted Tallamy's plan, but Bragdon objected. By linking gas tax income to bond repayment, the Tallamy plan limited the ability of the government to direct the pace of economic activity. Federal rather than state control of route selections was vital, he added, in order to assure priority of national over local interests. What Bragdon had in mind was toll financing of a 26,000-mile rural network, leaving to urban officials responsibility for building linking roads.[15]

Commissioner du Pont perceived Interstate construction in different terms. Few miles of Interstate construction could be made self-liquidating through tolls, he argued, and state engineers had constructed them already. Any plan adopted, or so it seemed to him, had to stress construction in urban areas, the points of major congestion.[16] In brief, the interagency meeting offered only a fresh forum into which government men extended their deadlock.

On October 28, after another month of haggling, officials at the Budget Bureau produced a proposal acceptable to Bragdon, to members of the Council of Economic Advisers, and to Secretary of the Treasury George M. Humphrey. In short, their plan combined Bragdon's national highway authority with the self-financing, flexible features on which Secretary Humphrey had been insisting. Officials of a National Road Authority (NRA) would organize and direct construction, leaving state highway engineers to provide administrative services. Bond sales and perhaps supplementary payment from gas tax income would finance construction. Forbidding a federal guarantee of bond repayment and excluding debts of the NRA from the national debt assured budgetary stability. By allowing the secretary of the Treasury Department to set the date of bond sales in line with the "requirements of monetary and fiscal policy," an additional check on economic gyrations was made available.[17]

But although officers of three departments had found some common ground, members of the Interagency Committee still were divided sharply. Disagreements went beyond finance and control, though they were vitally important, to a more fundamental conflict. Essentially, du Pont and other engineers such as Tallamy to whom he turned for advice were opting for a traffic-count version of highway construction. Bragdon and members of his small group defined road policy as part of grander plans for economic improvement and social control. Recent federal highway legislation was "only a start," according to Bragdon. He anticipated greatly increased employment opportunities, especially in the automobile industry, and acceleration of economic growth along new expressways. But all of this development, he added, would serve "as a continuous stabilizing force." Burns saw highway construction in much the same way. Accelerated road building, he believed, was a useful antirecessionary measure and would foster more efficient road transportation. Secretary Humphrey took an even wider view of a road program, stressing not only economic growth but perpetuation of the existing stratification system. Highways, he believed, were a "physical asset," and additional mileage would "create more and more" wealth for Americans. Rather than subdividing the fruits of production, he had told an audience of governors in April, 1954, it was preferable to "make another pie and everybody has a bigger piece."[18]

General Clay, however, was virtually indifferent to proposals sent his way by Treasury Department and Budget Bureau officials, producing a second rift within government circles. But by virtue of the president's instructions, or so they liked to think, Interagency members assumed a right to prior examination of Clay's proposals. If he would not cooperate, treasury officials threatened to withhold endorsement of his recommendations.[19]

Clay also had to contend with conflict between leaders in the road transport field. Farm leaders sought more mileage at less expense to their constituents, all without diminution of their own influence in local road-building affairs. Auto Club leaders argued for more attention to packed Interstate roads in urban areas, preferably by chopping farm–market construction from the federal payroll. Truckers, as always, wanted more roads built, provided only that taxes remained low. By October, then, the euphoria reported a few months earlier by Robinson Newcomb had degenerated into the usual bickering. According to one observer, "hearings which the [Clay] . . . Committee held . . . did not reveal any . . . consensus with respect to . . . finance." What it came down to was that "suggestions reflected . . . the interests of the group which the speaker represented."[20]

During October and November, 1954, Clay blocked out his own version of a remodeled highway program. Directors of a Federal Highway Corporation would handle financing of about $2.5 billion worth of construction yearly while the commissioner of public roads supervised operations. Executives of the corporation would issue bonds—about $25 billion worth—and retire them over thirty years with gas tax income and occasional borrowing from the treasury. Since traffic would increase, a point on which all agreed, an increase in the gas tax was unnecessary. So lucrative was this arrangement, figured Clay, that it would generate sufficient funds to pay 90 percent of Interstate system costs, to refund debts of toll authorities, and to bring their roads into the Interstate system.[21]

More than anything else, Clay had written a program close to standards set by Eisenhower and many governors. Bond sales and use of an authority guaranteed a stepped-up construction program and a strong stimulus to the economy, and motorists and truckers would finance everything without adding to the national debt or saddling the treasury with another burden. By accepting 90 percent of Interstate costs, Clay hoped state officials would "spend more money," thus "pump priming" the economy. Such reasoning corresponded with Eisenhower's larger interest in creating structures to move the economy along a steady upward course. "Our whole industrial activity," he wrote to Clay on January 26, 1955, had to be "geared to a purpose of steady and stable expansion." Fluctuations, or "the 'peak and valley' experience," he thought, "can make for us many serious and even unnecessary difficulties."[22]

Few in the federal government besides Clay himself, Eisenhower, and Adams approved each feature of Clay's plan. In late January, 1955, members of the Council of Economic Advisers wanted a "thorough go" at the report. They were thinking in terms of manipulating tax, toll, and bond rates and increasing the discretion of the national highway director. In short, they wanted a program with tools for economic management "built in." Secretary Humphrey, always interested in protecting federal revenues, insisted upon a clear-cut technique for directing gas tax receipts to the corporation.

His subordinates as well as executives at the Budget Bureau wanted profits from tolls—should any be imposed—to wind up in the Treasury Department. Such ongoing disputes managed to create enough confusion in administration circles to delay Eisenhower's road message to Congress, promised originally for delivery on January 27.[23]

But in meetings with Interagency Committee members, Clay remained adamant; modifications were bad politics. Any scheme promising direct federal aid for construction of toll roads would be " 'whipped' before it got started." If tolls were charged on previously free roads, there would be a "revolution" in several western states. Inclusion of other proposals from members of the interagency group would also jeopardize his program. As far as Clay was concerned, sufficient incentives were built into his program to guarantee support. Unless members of the interagency group affirmed his financing proposals, he preferred that they not forward his plan at all. On February 1, following strong urging from Sherman Adams, members of the Interagency Committee decided that Eisenhower's message to Congress and a subsequent administration highway bill would follow themes established in Clay's report. On February 22, Eisenhower proclaimed his support for a highway program along lines recommended by General Clay.[24]

Between January and March, 1955, highway users and builders as well as engineers—all men long anxious to get on with the right sort of highway program—endorsed Clay's proposal. By mid-1953, after Project Adequate Roads failed, they had decided to forget tax-free federal highway building and other difficult-to-achieve financing ideals. Governors of states with unusual congestion and high construction expenses looked forward to more help from Washington for costly expressways. Defense Department officers were most interested in completion of a compact, limited-access road system within a specified time period. For many business and professional leaders such as Pyke Johnson of the Automotive Safety Foundation, the Clay plan represented the culmination of more than a decade of promoting express highway construction at the state and federal levels.[25]

As part of an overall effort to win additional endorsements for Clay's program, President Eisenhower himself made several personal appeals. On February 16, he invited Clay to the White House to brief Senators William F. Knowland, H. Styles Bridges, and Eugene D. Millikan, and Congressmen Charles A. Halleck, Joseph W. Martin, and Leslie C. Arends. While Clay sketched his plan, Eisenhower limited his own role to asking leading questions, serving to direct Clay's attention into new areas. But the point of the meeting, insofar as the president was concerned, was to highlight for senior Republicans the urgency of constructing more roads in order to bolster the economy. "With our roads inadequate to handle an expanding industry," he told them, "the result will be inflation and a disrupted economy." Most of the airports built recently were obsolete already, he added, and "we cannot let that happen on our roads."[26]

On February 21, at the urging of Clay, Adams, and other administration leaders, Eisenhower conferred with ranking members of the Senate and House public works committees and roads subcommittees. Never, remarked Senator Chavez, had the president called all members of a committee to the White House to discuss domestic legislation. Soon, Eisenhower promised them, more than sixty million vehicles will jam our roads, "and we will have to build up our highways to meet that traffic." A ten-year road program, one fashioned along lines prepared by Clay, was "vitally essential for national defense," and would "help the steel and auto spare parts industry." Ultimately, then, an updated road program was "good for America."[27]

If those who stood to enhance their professional skills and reputations or fill their pocketbooks could live with most of Clay's package, others who would benefit little opposed it. For leaders of the Farm Bureau Federation and Farmers' Union, Clay's program was anathema. They had wanted the federal government to pay greater attention to farm roads, not freeze their funding for thirty years in order to pay off bonds. Some complained about interest charges on the bonds—it was money lost for construction—and about the decision to keep road bonds outside the national debt. Governors of states lacking traffic to support toll roads disliked the toll reimbursement feature. Finally, there was fear in some quarters that Clay had granted excessive power to directors of the Federal Highway Corporation.[28]

Senator Harry F. Byrd of Virginia, chairman of the Finance Committee and for many years a strong figure in the Senate, took up most of these gripes in a critique of the Clay plan published in January, 1955. Clay, he thought, had violated sound fiscal policy and had centered excessive authority in the corporation. Byrd recommended continuing the federal aid road program more or less as it was, thus eliminating interest charges in favor of more highway construction and retaining control of routing and standards in the states. Clay's scheme, he contended, would create "fiscal confusion and disorder," and raised the specter of the "iron hand of the federal bureaucracy."[29]

Word of dissatisfaction reached Eisenhower and his top men promptly. Between mid-February and early April, in reports from competent observers and in personal contacts with friendly legislators, administration leaders learned that their bill was in serious trouble. Senator Bridges, at his February 16 meeting with Clay and Eisenhower, had pointed to complaints "on the Hill" of "windfalls" in the form of reimbursement to some states for roads built already. At the February 21 conference, Senator Albert A. Gore of Tennessee criticized spending some $11 billion for interest on the bond issue. "That money should be spent on roads," he argued. On February 28, Commissioner du Pont wrote a colleague about "a great deal of political opposition." On March 30, Eisenhower learned that the comptroller general, his own appointee, had testified against Clay's bill before members of the Senate Subcommittee on Roads. The next day, Congressman McGregor, the ranking Republican on the House Subcommittee on Roads, sent along even gloomier news to Sherman Adams. As it was constituted, warned McGregor, Clay's bill did not stand a "ghost of a chance" in either chamber.[30]

Response within administration circles to disapproval of the program varied. Eisenhower, Clay, Adams, and du Pont stuck by main themes. For members of the Interagency Committee, however, criticism of Clay's proposals offered a unique opportunity to lobby for adoption of their own plan. As Bragdon was aware, Clay's bill was not popular, thus allowing them "to suggest again the clear-cut, sensible, simple way of doing this job."[31]

But neither Bragdon nor other disgruntled members of the Interagency Committee had much to add. Again, Bragdon proposed creation of a national highway authority with broad powers, leaving state road engineers with supervisory tasks. In general, he still was hopeful that federal officials would gain another lever in economic affairs. Treasury Department officers, however, remained obsessed with toll financing. Seventy-five percent of Interstate construction could be financed with toll collections, or so one of them calculated.[32]

During April, 1955, Bragdon and his cohorts tried to develop a broader base of support for their own proposals. Yet since each had something different in mind, each moved in an independent direction. Treasury Secretary Humphrey attempted privately to bring Senator Byrd over to the side of the toll road forces. Bragdon, for his part, thought the most fruitful route lay through the White House. He presented his views to several of Eisenhower's aides and repeatedly urged Arthur Burns to bring them before Sherman Adams. Burns, however, viewed such activity as premature. Best, he thought, to contact other members of the Council of Economic Advisers and to discuss the situation with members of the Advisory Committee on Economic Growth and Stability.[33]

By early May, men of vastly different convictions about the direction of road building disliked parts of Clay's program. Constituents of the original road networks protested Clay's indifference to their highway needs; road users objected to making reimbursement for toll roads constructed in the future; still others disliked what they perceived as excessive power lodged in the national highway authority, a point often rooted more in self-interest than abstract principle. Army officials feared that creation of an authority would encourage finance men to dictate "engineering standards."[34] Members of the interagency group finally promoted a separate program, utilizing uncertainty created by criticism of Clay's bill to call for increased financing of toll roads and even greater centralization of control.

General Clay still opposed making any changes. The Interstate system created "a profit," he told a group of governors on May 2, and it was unfair to levy additional charges on its users in order to boost construction of low-volume roads. Besides, Clay perceived no reason to rewrite his bill. While it would encounter difficulty in the Senate, Clay believed it would receive a more positive response in the House. At this point, he argued, it was not a "lost cause in any sense of the word."[35]

Clay badly misjudged the extent and intensity of support for his program. By a vote of 8 to 4, members of the Senate Public Works Committee turned down the Clay plan, substituting a bill sponsored by the chairman of the Roads Subcommittee, Senator Gore. Basically, he simply raised the funding levels and time frame for federal highway construction, leaving administration and distribution of funds as before. Like Clay, he did boost the federal share of the Interstate construction bill from 60 to about 90 percent, but his only substantial departure from past practice was inclusion of the Davis-Bacon Amendment. It was an old federal law requiring payment of prevailing wages on federally sponsored projects, and Gore made it applicable to the Interstate system as a way of enticing leaders of organized labor. They had been largely indifferent to highway legislation, asking only that the Davis-Bacon Amendment be made a part of whatever bill was approved.[36]

Between May 20 and May 25, members of the Senate modified sections of Gore's bill. By a voice vote, for example, they dropped the Davis-Bacon section. But true to predictions, on May 25 they approved Gore's bill, turning down by a two-to-one margin Senator Edward Martin's motion to substitute Clay's measure.[37]

On June 28, House Roads Subcommittee Chairman George H. Fallon of Maryland introduced a bill designed to bring sufficient benefits to everyone. Somehow, he had to find a formula which provided stepped-up construction of Interstate and farm roads, left management alone, and did not disturb the federal budget, all without burdening truckers with huge tax increases. It proved an impossible task.

Higher taxes were the key to Fallon's thinking. By boosting gasoline taxes and automobile and truck excises on a graduated basis, he could promise completion of the Interstate system in twelve years and expansion of the rural road program. Taxes on the largest truck tires would jump from five to fifty cents a pound while gasoline and diesel fuel imposts would rise to three and six cents a gallon respectively. At those rates, income and expenses were supposed to balance in a number of years.[38] The budget would remain secure.

No one was very enthusiastic about Fallon's scheme. Graduated taxes and especially the prospect of a tenfold increase on big tires distressed truckers. Since western operators ran more diesel units, they were alarmed at the difference between diesel and gas tax rates. In order to construct needed roads, major truckers were willing to pay slightly higher taxes, but at a uniform rate.[39] In reality, they preferred the regressive feature built into a uniform tax schedule.

Administration officials were divided in their response to Fallon's bill. Of course, whatever measure Congress approved had to avoid deficit financing. Secretary Humphrey, then, preferred the Clay plan, relative to the Gore plan at least, yet endorsed Fallon's proposal insofar as it provided for a balance of income and expenses. But at his June 29 press conference, Eisenhower again spoke for the Clay plan, because both the budget and state finances appeared safer. "I am for it now," he told reporters, "just as strongly as I was when it was devised by the Governors and by the Clay Committee. . . ." But whatever they thought about one plan or another, as late as July 1 the feeling persisted within administration circles that members of the House would approve road legislation.[40]

On July 6, leaders of the House Public Works Committee appointed a special nine-member group and charged them to produce an acceptable formula. They were as unlucky as Fallon. Although committeemen were enthusiastic about the fruits of highway construction, looking forward along with most Americans to faster traffic and an expanding economy, they produced only a scaled-down version of Fallon's original proposal. In brief, their modified plan still included a graduated tax schedule on tires and an extra increase on diesel fuel. But Eisenhower stood by the Clay bill; and nearly five hundred truckers went to Washington to complain about the inequity of graduated taxes.[41]

Support for all highway legislation disintegrated. Better no bill, reasoned many in and out of Congress, to one appearing to violate their own standards of economic decency. House debate and voting patterns followed these sentiments. On July 26 and July 27, House supporters of the Clay plan offered a couple of compromises, dropping toll road repayment and adding the Davis-Bacon Amendment. Congressman Charles Halleck suggested financing an expanded road program with bonds for twenty rather than thirty years by raising the gas tax slightly; but his motion was ruled out of order. At this juncture, one of the proponents of Clay's bill called on his colleagues to "stand by the President of the United States and support the President's program."[42]

With lines drawn so taut, every alternative was struck down. On July 27, by a vote of 193–221, House members rejected a motion to recommit Fallon's modified bill and substitute Clay's. Those who had voted for Clay's bill then joined the opponents of Fallon's to kill it, 123–292. On August 2, Congress adjourned for the year, its members split over the course of the American highway program.[43]

Administration leaders discussed calling a special session of Congress to deal with highway legislation. Burns, in particular, urged Eisenhower to do so. For his part, the president pleaded with Congress to reconsider, but by August 4, he had decided not to require a special session. Reconvening Congress, he thought, "could be at the cost of the sanity of one man named Eisenhower." In 1955, then, hopes of reforming the federal aid highway program were dead.[44]

Between 1939 and July, 1955, everyone with a stake in highway legislation had tried his hand at revamping the federal road program. Certainly, Eisenhower's reform effort had been the most strenuous undertaken by a president. Although he and his aides had succeeded in narrowing the parameters of debate, they could not produce a plan capable of breaking through the deadlock.

Failure of Bragdon and members of the Interagency Committee to achieve their minimum objectives is easiest to understand. According to Eisenhower's arrangement of committee responsibilities, interagency members were supposed to consolidate their own thoughts and present them to Clay. From the beginning, then, members of the interagency group stood below the quasi-governmental Clay group, a structural fact making their proposal only one of many brought before General Clay. In a sense, they too were just another "special interest." Ultimately, dissatisfied members of the Interagency Committee enjoyed few options except to lobby with Clay for revisions, affirm his decisions, and then push surreptitiously for changes as the program wallowed in conflict.

Had committee relationships been structured otherwise, it is not certain that interagency members would have done much better in the national political arena. Key features in their own program, when finally written, had themselves long been unpopular with important political, professional, and business figures. Bragdon, Burns, and Humphrey dreamed of creating a national road authority, one controlling highway construction in the interest of economic growth and efficiency and economic and social stability. Their plan, if ever implemented, aimed to revolutionize the political economy of road building, mostly a decentralized affair not usually related instrumentally to economic movements. While the federal government paid about 50 percent of the expenses of building roads on the federal aid system (about 60 percent for Interstate), county and state engineers made design, routing, and construction decisions, only needing federal certification of their plans and work. For their own part, engineers aimed to serve traffic by widening or rerouting older highways and by constructing new ones to areas not well-served. Most truckers and nearly every governor, engineer, highway contractor, and congressman liked it that way.

But finally, legislative ineptitude killed the Clay plan. Truckers and everyone else wanted more roads, but members of competing groups entertained strong preferences about finance and administration, opinions rooted in local commercial and professional advantage and traditions of their own industry. Bragdon, however, characterized alternatives such as Gore's plan as a "headline bill" while James C. Hagerty, Eisenhower's press secretary, defined it as "the old Democratic arguments."[45] Since these images were prepared for the eyes of administration officials alone, they suggest a view of opponents as men seeking narrow partisan advantage. Beginning in February, reports that Clay's bill would not survive in Congress—brought in by men with years of experience counting votes—were discounted. Surely, Eisenhower's aides must have reasoned, with such strong support from state governors, men who were playing politics would succumb to political pressure. Thus, Eisenhower, Clay, and Adams, also men with strong convictions about good highway legislation, stuck to their original plan. All, then, preferred to sacrifice highway legislation.

However anxious were organized road users, legislators, and members of Congress for new highways, debate at the national level about the direction of the national road program continued for another year. Only in mid-1956, when Congressmen Fallon and Hale Boggs found a formula which satisfied basic concerns and demanded few sacrifices, could administration, congressional, and industry leaders agree to go ahead with an accelerated road progrm. Until then, irritated motorists, cost-conscious truckers, and model-building economists and administrators had to wait.

7

The Interstate Highway Act of 1956

If it were not for the urgent need to get the big highway building program under way without further delay, every red-blooded trucker and his legion of allied industry and shipper friends would switch his position from vigorous support of the highway program to an out-right, last-ditch battle against the entire program. Unfortunately, that is what the railroads want the truckers to do so that the truckers would be blamed for killing the highway measure which the scheming railroads had set out to do by "hook or crook." The trucking industry instead is looking to the Senate committees to restore equity and reality to the tax increase measure.

William Noorlag, Jr.,
General Manager,
Central Motor Freight Association,
March, 1956

Defeat of all road legislation did not soften the opinions of competing highwaymen and political leaders. Beginning in August, 1955, they lobbied for their version of good highway programming, once more debating the virtues of national control of road construction, the merits of toll and free highways, and the proper rate of gasoline taxation. So troublesome were these matters, many believed that Congress would hold up legislation again. In June, 1956, however, members of Congress voted overwhelmingly for a bill fashioned by Congressmen George Fallon and Hale Boggs. Actually, neither Boggs nor Fallon had much new to offer. In brief, they wrote a bill providing competitors with nearly everything, all without asking extraordinary sacrifices of principle, practice, or cash from any group.

General Bragdon remained the most diligent proponent of toll financing. Beginning on July 28, 1955, only one day after House members had rejected any change in road-building arrangements, he claimed that toll collections would finance at least 23,000 expressway miles. Provided earnings were transferred from one state to another, Bragdon estimated that about 30,000 miles could be financed from tolls.[1] On September 27, Bragdon outlined his plan in a letter to Sherman Adams. Somehow, he promised political support from those who opposed tolls, and pointed again to "the great savings to the taxpayer." Senator Prescott S. Bush, Representative Jesse P. Wolcott, and others favored toll financing, or so he claimed. Now was the time, then, to "prepare a bill incorporating these factors."[2]

But Bragdon's schemes and proposals did not impress Adams or other top officials, assuming Adams even bothered to call them to their attention. At the cabinet meeting of September 30, while Eisenhower was hospitalized with a heart attack, senior officials formed a Cabinet Committee on the highway program. The secretary of commerce, Sinclair Weeks, would serve as chair; Secretary Humphrey, along with the secretaries of defense, agriculture, and labor, and a representative from the White House Office comprised the membership. Before November 3, the date set for a meeting with state governors, they agreed to review Clay's program and congressional action during the spring and to recommend modifications, if any.[3] Since men with considerably more leverage had not been able to break the deadlock, about all they could do was gather up statements of opinion, aspiration, and hope from government and business leaders and try to set the outlines of an acceptable bill, much as Clay had done before. By October 1, another group of Eisenhower's administrators had gotten a search for a new federal road formula under way.

Results were only a little better than before. In road politics, there were no secrets. Truckers had made public, usually often, what they expected. At a series of conferences held during the last two weeks of October with members of the Cabinet Committee and their aides, heads of the trucking industry told their story again. Bonds and administration and anything else did not matter, just tax rates. Because the Fallon bill imposed differential rates, especially on tires, they had opposed it. Truckers, a leader of the American Trucking Associations claimed, "were singled out in the Fallon Bill as the whipping boys." Tax equity, as they figured it out, amounted to uniform, one or two cent hikes on gasoline and tires. Without objection, moreover, they would pay another 2 percent excise on new trucks, provided proceeds went straight to highway construction. Tollways were a different matter. Traffic on toll roads moved easier, and truckers liked that well enough, but they opposed schemes encouraging further toll collections or extensions of toll networks.[4] As other Americans, truckers preferred low taxes; as their trucking forefathers, they preferred regressive ones.

Word from Eisenhower's administrators ran along different lines, but echoed old claims and notions. Rates, as such, did not matter as long as road construction financed itself; so much the better if they got another handle on economic development. Each man, however, emphasized one or the other. Treasury Secretary Humphrey, always an independent actor on the highway scene, concentrated on ensuring self-sufficiency for any road project. He preferred toll financing, but would accept an earmarked user tax if equal to expenses. General Bragdon, by way of contrast, remained anxious to build stabilizers into the economy. On October 18, he pushed members of his Committee on Public Works—another federal interagency group—to prepare long-range construction plans. While such ideas appealed in principle, most on the committee sought immediate plans and fewer regulations, one participant asking if during an economic emergency he "could expect relief from detailed restrictions on contract procedures." Members of the Council of Economic Advisers also focused on linking road expenditures with economic fluctuations. And on November 1, Gabriel Hauge, Eisenhower's personal economic adviser, wrote Secretary of Commerce Weeks to ask if his Interstate system plans provided mechanisms for directing the economy. "That was the fundamental purpose of the plan in the initial instance," as he recalled it.[5] All in all, in administration circles, defeat had not dampened enthusiasm for drastic changes in the federal road program.

By the end of October, members of the Cabinet Committee had fashioned the outlines of a highway bill, just as promised for their upcoming meeting with state governors. After another month of letters, conferences, and memos, all they suggested was that the administration offer the same program as before, adding "a cumulative increment annually" for construction of non-Interstate roads, dropping the road-building authority. Governors would not endorse an increase in federal gasoline taxes, or so cabinet officials thought, but they recommended a boost "as may be necessary after deducting other federal aid." Members of the cabinet planned no deficits. Secretary Humphrey, however, believed that taxes sufficient to finish the program within ten years would not be raised. Construction time, he feared, would drag along for sixteen to eighteen years.[6]

Whatever Eisenhower's economists and cabinet officers or even the governors themselves thought, few in Congress, the road transport industry, or the Council of Economic Advisers paid much attention. During November and December, transport and road-minded men continued to plan a highway program to taste. In correspondence dated November 4 with a White House aide, Congressman McGregor argued for dropping the Interstate system from future legislation. Within the week, however, Congressman Fred Schwengel of Iowa's first district wrote a colleague to propose financing all construction with bonds and increased taxes on tires. But during November and early December, General Bragdon—himself well aware of Cabinet recommendations—corresponded with bureau engineers about another of his toll finance schemes. And as late as December 13, members of the Council of Economic Advisers, men thinking in still broader terms, remained committed to bond financing, to creation of an independent authority, and to legislation aimed at "dovetailing of construction expenditures with general economic conditions."[7]

If men continued to plan for their own version of an upgraded federal road program, most of them, whether in or out of government, were pessimistic about the chances of anything passing Congress. Prospects appeared dismal, reported long-time observers of highway politics. At the November 17 meeting of the "Road Gang," an informal group of highway users and contractors based in Washington, D.C., a leader of the Highway Users Conference claimed that "any highway plan having a built-in financial plan" was in for a rough "go." Highway legislation was blocked, claimed an auto industry executive at the gathering, by "political times."[8] Late in 1955, as in debates past, all wanted more highways. Each, however, recognized the limits of his own willingness to compromise and the downright intractability of opponents.

On April 27, only four months after deadlock appeared certain, members of the House approved a major highway bill sponsored by Congressmen Hale Boggs of Louisiana and George Fallon of Maryland. Fallon contributed the details of fund distribution, control of construction, and apportionment; Boggs concentrated on finances. The vote was 388–19.[9] Beginning soon, if the Senate concurred, Interstate, farm–market, urban, and main trunk road construction would enjoy a gigantic boost in federal aid.

In speeches on the floor, congressional leaders ascribed their success to a sense of compromise and moderation, both among members of Congress and those in the road transport and construction industries. Major differences, concluded one legislator, were "at the threshold of being resolved." Congressman McGregor, usually a sturdy proponent of farm–market road building, thought this consensus was the "result of study and 'give and take.' "[10]

In part, appraisals by House members were accurate. Give and take did occur. During January, 1956, Secretary Humphrey and President Eisenhower had talked on the telephone of exercising greater direction over economic movements. Eisenhower himself was seeking some technique to get top corporation executives "to listen to us in advance hereafter." By the end of the month, however, both were willing to forgo such discretion in the highway field. Following a January 31 meeting with congressional leaders, Eisenhower's aides were told to "yield to Democratic insistence on financing" and to "cooperate in the development of an appropriate tax proposal." Senator Byrd, chairman of the Finance Committee and critic of the Clay program, was to "be consulted as to the most desirable procedures for expediting the bill."[11]

In the trucking industry, too, men talked of concessions. Between February and April, 1956, in meetings, in industry-wide publications, and in correspondence with members of Congress, they announced again their willingness to pay higher taxes. They complained about registration fees. But a one cent a gallon hike on gas and diesel, three more cents a pound on rubber, and a 2 percent additional excise on new vehicles, all as Boggs was recommending, appeared endurable. Beginning as early as February, then, once rates were fixed, truckers urged House members to vote for the Boggs-Fallon bill.[12]

Actually, few in government or industry had made major concessions; fewer acted from some spirit of give and take. References to compromise, at least in highway politics, served as functional myth. The fact of the matter was that Boggs and Fallon had written legislation which incorporated long-sought goals, asked few significant sacrifices, and managed to sidestep difficult questions.

Basically, the key to success was providing something for everyone without imposing high taxes on truckers. Distribution of funds —at first for farm, urban, and trunk roads, later for Interstate routes —had been a sticky issue since the beginning of the Federal Aid Highway System. Fallon handed out record high sums for each, and promised another $25 million yearly for urban as well as rural construction. Urban supporters of Interstate construction came out best. The federal government would pay 90 percent of Interstate expenses, about $25 billion, but distribute the money according to local needs. Since costs in congested urban areas were greatest, they would receive a disproportionate share of funds. In order to finance all this construction, Fallon and Boggs increased automotive taxes. But they largely went along with truckers, asking for moderate, ungraduated increases. Only Representative Daniel A. Reed's amendment, one imposing a surcharge of $1.50 per thousand pounds on the total weight of trucks heavier than 26,000 pounds, appeared out of line with their willingness to pay.[13]

Delaying action on divisive items was the second factor in the success of the Boggs-Fallon bills. For years, differences of opinion about toll road repayment and the even more vexing matter of tax equity between big and small truckers, bus operators, and motorists had convulsed road politics. Boggs and Fallon avoided both. Because most agreed that the federal government should pay compensation to states for toll and free roads built already and added to the Interstate system, Boggs and Fallon were able to make that promise. But they delayed a decision on which roads were entitled to a credit pending a study of standards by the secretary of commerce. In turn, Congress would review the results of the secretary's study. No doubt, a lengthy study ordered by Boggs and Fallon of road costs assignable to auto, bus, and truck operators was intended to set aside that question too.[14]

Boggs and Fallon also prescribed industrial and professional standards for highway finance and construction. Since the 1930s, leaders of auto and trucking associations and state road engineers had complained that governments collected more in motor vehicle taxes than they spent for roads. State legislators had crystallized these views into antidiversion and trust fund arrangements. Boggs and Fallon found a place for the antidiversion impulse by creating the Highway Trust Fund. Revenues from taxes on fuels, tires, and new vehicles and Reed's surcharge would go directly into the Trust Fund for road building alone. Finally, Boggs and Fallon allowed advance condemnation and limited access design.[15] No longer would engineers and users suffer intolerable delays in acquiring land; no longer would they have to endure the nuisance and hazards of cross traffic.

If the Senate was going to pass a road bill, as now seemed likely, then nearly every leader of government and industry associated with road transport had something special to include, something equally vital to get dropped. Inclusion of the Davis-Bacon Amendment, requiring payment of prevailing wages as determined by the secretary of labor, excited the most controversy. Contractors and state road engineers worked hard to eliminate the amendment. As early as January 19, members of a group of engineers and contractors had declared for local determination of wages, invoking mostly cost arguments. Beginning around March 1 through early June, contractors and chamber of commerce officials joined the struggle against Davis-Bacon, sending letters and petitions to members of Congress. Usually, they spoke of efficiency, of lower costs, of states' rights, all symbols, images, and commercial realities celebrated by men in contract road work. In April, administration leaders took up the anti–Davis-Bacon cause, trying to find a way to cut it without angering labor leaders.[16]

Then, too, there were the questions of truck taxation and a balanced budget, each the personal crusade of one or two top officials, each also to be resolved by Senator Byrd. On March 19, Senator Lyndon B. Johnson, the majority leader, wrote Byrd to secure elimination of the first thirteen tons of truck weight from the surcharge imposed by the House. For their part, treasury officers focused on the budget. While the trust fund was supposed to balance after several years, at times expenses would exceed income, requiring the treasury to make up the difference. On March 23, Secretary Humphrey asked Byrd "to do something about this in the bill." Loans, he argued, "will put us in trouble in the general budget." Humphrey also thought it vital to include the surcharge, the one "to which Lyndon Johnson objects," or "the figures would be out by that much more." All in all, apparently, now it was a matter of looking after more basic concerns, more specialized interests, more basic animosities.[17]

Members of the Senate had to deal with all these troublesome matters plus the usual disagreements over who got what. In brief, on May 29, they opted for a road program aimed at pleasing commercial highway users, rural road enthusiasts, and key members of the Senate and administration. They accepted most of Representative Boggs' financing measure, the one truckers liked, and added the Byrd-Humphrey Amendment prohibiting trust fund deficits. Following Johnson's lead, they exempted the first thirteen tons of truck weight from the surcharge. Apportionment, as before, would favor construction on the old line networks; careful calculation by many of state receipts under alternative schedules apparently suggested the wisdom of voting again for the tried and true. None of this agitated anyone. Only Davis-Bacon provoked senators to real controversy. At one point, so great the confusion, so diverse the approaches, the Senate voted Davis-Bacon or a revision of it several times, achieving that many different results.[18]

The forthcoming meeting of Senate and House conferees served as another arena for old competitors to lobby for some local interest, some national need. On June 1, Senator Stennis wrote to Senator Chavez to urge him to "hold out" against efforts to cut funds for farm–market and main trunk roads. Should Interstate construction win too much support, he warned, it will be like the "neglected calf" who was "knocked in the head with the churn-dasher." Through most of June, truck operators, with their own commercial needs to look after, sent telegrams to Chavez opposing size and weight provisions; administration leaders focused on finance, particularly ensuring endorsement of the Byrd-Humphrey Amendment; and Mayor Robert F. Wagner of New York asked that tenants receive aid to help defray the costs of moving. In 1956, of officials who even thought of such matters, Wagner was one of the few who favored compensation.[19]

Although major issues were settled, Senate and House bills differed in significant respects, particularly apportionment. Then, too, important men in government were demanding greater and lesser adjustments. Negotiations between conferees were difficult; at one point they considered returning without a bill. On major items such as fund distribution, however, they compromised more or less. Between 1957 and 1959, Interstate money would be distributed according to the Senate formula: so much to each state based on land, population, and road mileage, just as always. For the remainder of the program, between 1960 and 1969, states would get their share of Interstate costs as a percentage of total Interstate costs, thus hopefully ensuring timely and uniform completion. In areas such as the Byrd-Humphrey Amendment, House conferees gave way. As part of the exchange, if that is what they were doing, conferees eliminated provision for 1,500 miles from the Senate version, leaving the Interstate at 41,000. They also amended the name of the Interstate system, since 1944 itself an engineering and legislative fixture. As of 1956, it became the National System of Interstate and Defense Highways. But the overall emphasis still was on federal financing (roughly 90 percent), rapid completion of the Interstate system, and continuing modernization of the rural road networks. On June 25, conferees submitted their report. On the twenty-sixth, members of the Senate approved it, 89–1; members of the House, on the same date, approved it, but did not bother recording their vote. On June 29, President Eisenhower signed the bill.[20]

After nearly fifty years of traffic jams and even more of urban decay, after a quarter of a century of fumbling efforts to use road construction to direct economic activity, leading parties had agreed to an accelerated highway building program. Boggs and Fallon had found the key to success. They promised plenty of new roadway for everyone and security for treasury deposits, and had asked truckers to pay only modest tax increases. At the core of this formula was the decision of truckers and leaders of motorist associations, however reluctant, to sponsor the entire federal aid highway program. Once financing was arranged, congressmen were left with the relatively easier task of imposing professional standards on federal road projects and spreading revenues among competitors.

Additional developments, in politics, in the economy, in their industry, encouraged truck operators, engineers, and political leaders to write an agreeable road bill. In the first place, 1956 was to be the year of decision, or so many thought. If national leaders could not get highway construction rolling, then local officials, those more subject to particularistic influences, threatened to launch their own road programs. Late in December, 1955, Cleveland officials met with a federal road engineer. Including a block payment from the state, Cleveland had funds available to complete only a few sections of their expressway network, and now they had begun to explore with him the prospects for toll financing. Others launched a renewed attack on federal collection of gasoline taxes. "If Congress does not enact a highway program bill this year," a South Carolina engineer told delegates to the AASHO meeting on December 6, 1955, "the federal government should withdraw from the field of special taxes on motor vehicles and let the states pick up this revenue."[21] Local construction, however, raised the specter of higher costs and hated toll fees, and would mean a disjointed rather than a uniform effort.

More general notions, more general dreams, dissolved some of the remaining tensions. Americans, or at least those who wrote, argued, and made decisions about highway building and traffic matters, were optimistic about the natural congeniality of highway construction and economic growth. If traffic tangles were reduced, if billions were spent for more roads, the economy would prosper. Truckers and contractors, then, could look forward to personal wealth; economists and government officials, to steady economic growth; and farmers and urban motorists alike, to faster trips to market, to jobs, and to recreation areas.[22]

American commitment to automobility—the conviction of most that motor vehicles and fast-flowing expressways were good in their own right—also facilitated efforts to find a solution to the legislative tangle. For several decades, motorists had been stalled in traffic jams. Midway through the 1950s, many had determined that highway mileage pure and simple was more important than apportionment and finance formulas. By January, 1956, according to a publicist for the AAA, motorists wanted "better highways now." President Eisenhower certainly saw things that way too. Initially, he had insisted upon the Clay plan. After losing that battle, however, Eisenhower was ready to sign any bill as long as it included a self-financing feature. In 1956, the president "just wanted the job done."[23]

8

Highways and the Values of Americans

The Federal-Aid Highway Act of 1956 foreclosed most of the options in American road politics. For the next two decades or so, engineers busied themselves completing the Interstate system, still immersed in squabbles about farm and urban road funding, economic growth, social control, and urban development. Under 1956 legislation, highway spending was inflexible. But during October, 1956, Burns and Bragdon were still seeking some method of varying the pace of road expenditures, hoping to develop another countercyclical weapon for the administration. Because "the horse . . . was out of the stable," Bragdon recognized that it would prove difficult to revamp the act. After some indecisiveness, however, Bragdon and Burns fixed on the idea of advancing an amendment allowing deficit finance in the event of recessions, limiting road spending under inflationary conditions. If that tactic failed, they looked forward to reviewing financing procedures during a general examination of road finance scheduled for 1959. Chairman Burns was not concerned with the political feasibility of his proposals, just "their technical and administrative soundness."[1] Neither initiative was successful.

Between 1958 and 1960, although Arthur Burns had left the council, Bragdon pursued old dreams. In December, 1958, and again in March, 1960, he sought a hearing at Cabinet meetings for his views on toll finance, more centralized construction, and limiting Interstate construction to rural areas. Toll finance in particular appeared vital, he later reported, "if this juggernaut of tax consumption is to be checked and not cancerously devour funds that other needs have equal rights to." Finally, on April 8, 1960, Eisenhower met with Bragdon and several top officials. But according to notes of this conversation, the president had concluded that Interstate construction "had reached the point where his hands were virtually tied." By June, 1961, in Bragdon's flattering self-portrait, he was "a voice in the wilderness."[2]

The picture was much the same in terms of utilizing Interstate roads for urban redevelopment purposes. Since federal and state road engineers controlled the program, they had few incentives to include urban renewal, social regeneration, and broader transportation objectives in their programming. Their task, as they saw it, was one of promoting traffic efficiency by constructing roads. By 1958, so intense was criticism from planners of the way in which engineers were handling urban road construction, leaders of both gathered for a week-long meeting at the Sagamore Conference Center in the Adirondack Mountains. The Sagamore Conference, reported the head of AASHO several months later, marked "a beginning place for cooperation in urban development that has never existed before." But day-to-day relationships and priorities did not change. While the 1962 Highway Act mandated consideration of urban transportation as well as the city as a holistic package, engineers were able to maneuver around it adroitly, building roads largely as they wished. Only where urban leaders were committed to a unified program of renewal were expressways built as part of some wider plan. Basically, then, traffic patterns of motorists and truckers and decisions of engineers determined the outlines of Interstate construction.[3]

Actually, post-1956 highway politics was only a scaled-down version of earlier developments. Indeed, conflict had been part of the highway construction scene for years. Certainly, participants shared a number of goals and ideals. All, for instance, celebrated low taxes, balanced budgets, and traffic efficiency, and each also thought that additional federal aid should not interfere with private choices, with independent action. More roads, finally, seemed vital to economic growth, perhaps even to social harmony. While this potpourri of conflicting notions appeared coherent and equitable, highway politics had been deadlocked since the depression, leaving the pace of road construction well behind the appetite of motorists.

Depression and inflation, factors outside personal control and professional meddling, limited road mileage somewhat, but the real problems were social in nature. Diverse professional, bureaucratic, and commercial experiences—in construction firms, in engineering offices, in the different opportunities of large- and small-scale truckers—shaped vastly different perspectives on mounting traffic, economic, and urban problems. That road construction should serve traffic alone was a principle on which major truckers and engineers stood together, and both worked tirelessly to stop diversion and unwise spending on farm roads. Whatever the theoretical justification, and publicists turned out plenty of them, in practice truckers and engineers objected to make-work road building and refused to pay for expressways routed to restore some larger urban complex. Roads, whether in 1938 or in 1956, were going to be built for commerce.

Advocates of city remodeling, including professional planners and a few urban businessmen, made up another element in expressway politics. Professional training and experiences, in the case of planner Harland Bartholomew, encouraged him to view urban areas as a whole. For corporate leaders such as Richard Mellon, investments and responsibilities in national firms suggested the importance of city-wide and even regional renewal efforts. Broad-scale renewal was an effort to make urban areas prosperous and urban affairs congruent with their own outlooks and business needs. These men, both planners and business executives, accepted norms of traffic efficiency and idealized economic growth, but they hoped to channel them according to expansive definitions of social development. Expressway construction, many of them thought, should assist with local renewal projects, maybe serve as part of an urban transport system including mass transit and additional parking places. Occasionally, they spoke of renovated communities, of new regional centers. What they envisioned, more than anything else, was government sponsorship of an elaborate and costly rebuilding program, one which would supply social overhead for sustained and moderate economic growth. Federal transfer payments, in this scheme, replaced risky private investments in downtown areas, ensured the safety of others, and left direction of local affairs in the hands of local leaders.

President Roosevelt and then Presidents Truman and Eisenhower and their top aides looked forward to faster traffic, often as well to government promotion of economic development. Yet they were at odds with engineers and road users. All three presidents defined their own job as one of national economic management, not serving any one industry such as trucking and road construction, and each sought special mechanisms to speed up or retard road construction in order to foster sustained economic growth without further upsetting their budgets. Between 1939 and 1954, they argued in favor of relatively low road outlays. Then about a year through his first administration, Eisenhower and his economists endorsed a big jump in road spending. They hoped to relieve traffic congestion and to hitch the economy, in part, to a gigantic public works project, one financed and structured to safeguard the budget.

By 1955, although differences remained, all the perceptions and commitments needed for a stepped-up road program were in order. While participants in road politics did not recognize it, they had settled most of the outstanding questions. In 1955, they lost road legislation in Congress because of differences over the details of finance. By 1956, the press for more roads and a bill which asked few sacrifices, especially from major truck operators, dissolved these differences.

The 1956 Highway Act, which authorized stepped-up construction of the national expressway system and hearty and regular increases in aid for building urban, primary, and farm–market roads, emerged from this social and political milieu. It was a highway-building program pure and simple, one which federalized state and local road-building practices and ideals, one which preserved standards and arrangements long presumed just and normal by engineers, truckers, congressmen, and governors. In 1956, just as in 1944, proponents of industry practices, of business autonomy, of local initiatives, and of governance by established routines in trucking and engineering circles had managed to block those who were driven by an urge to plan American transport and American society in detail. Only, of course, when state and county officials had proved incapable of locating funds for farm–market, urban, primary, and Interstate road building did truckers and other highway enthusiasts themselves even turn to the federal government for massive highway spending. What they managed to secure, then, was federal funding for localistic and largely impermeable commercial and professional subcultures. Nevertheless, after 1956, just as before, city planners served as the cartographers of urban expansion; economists kept exacting and complex charts of economic bumps, of private investment decisions, of fixed federal outlays.

Alternatives, in the form of plans for highway building as part of regional centers, downtown renovation, and controlled economic expansion, were available as antidotes to concentration on expressway construction. Beginning during the 1890s, city planners wrote of extra wide rights-of-way; by the 1930s, they had constructed greenbelt communities. In 1939, GM's Futurama capped this tradition, promising highways which would regenerate urban centers, boost farm income, and produce a traffic utopia. During the 1940s, members of the National Resources Planning Board and their researchers such as Wilfred Owen spoke of expressways as part of their vision of uniform transportation planning and comprehensive urban development. Toll charges and a national transport agency, as they liked to think, would encourage development of air, water, and rail services. By the mid-1950s, Charles Dearing and Robert Murray, Jr., in the Department of Commerce were opting for toll charges to foster more uniform allocation of transport facilities.

But these plans and ambitions were not feasible politically. In the first place, those who favored comprehensive transport and urban planning were as divided as supporters of highway building alone. Toll road advocates such as Bragdon were centralizers in terms of road building and economic management, but perceived the fruits of their efforts contributing to unlimited economic development. Surely, city planners and allied business executives liked growth well enough, but each advertised his own work as a vital contribution to controlled expansion, to enhanced leverage over local social patterns. But as men in city halls, county and state road departments, and trucking company headquarters planned their own futures and their own investments, talk of toll roads and regenerated cities appeared expensive, unproductive, and maybe even dangerous.

Unity among those who promoted transport and urban alternatives would not have mattered anyhow. Since 1921, road engineers had controlled state and federal highway construction and finance and along with truckers had opted for road building, not schemes for social and economic remodeling. Most Americans looked at highway construction in these terms as well. They usually voted to limit gas tax revenues to road construction, keeping them from the general funds and diversion to other state services.[4] Members of the Bragdon group, including Secretary of the Treasury Humphrey, tried to break this pattern. Although they were often divided, inept, and obsequious, they managed to produce a thoughtful plan for toll finance and centralized management of highway building. But it was a radical proposal in most respects, one which threatened destruction of established methods of federal-state road building and finance. If their proposal had been implemented, perhaps Clay's prediction of "revolution" would have proved accurate.

Deadlock and conflict in road affairs were part of the larger American social scene. While competition for a bigger cut of national income was at the heart of many disputes, increasingly narrow definitions of job responsibilities widened the social gulf in every industry, in most cities. Business and professional leaders, when it seemed appropriate, turned to trade groups, to professional organizations, later to government agencies for aid in crushing competitors and blocking unfavorable action by others. All the while, huge bureaucracies emerged to handle day-to-day problems of a complex social order. But federal bureaucrats, say for instance Thomas MacDonald, soon turned their attention to gathering greater influence over the affairs of county and township officials. In turn, localites resisted encroachments with strength and skill, in good measure by creating national agencies staffed by their own bureaucrats.

One of the key factors in the development of these disputes was the lack of external controls from government and trade associations and no more than formal adherence by most Americans to unifying norms and values. Talk of privacy, of private property, of individualism, always won a consensus, and most Americans genuflected before the ideals of thrift, diligence, and equality of opportunity. But political parties and government agencies as well as business and professional leaders reflected a particular and often a local view of things. Their perceptions of national needs were rooted in narrow definitions of industrial and professional welfare and the exigencies of bureaucratic survival. As a result, they were restricted in their ability to determine common values. In a fragmented society, there were few who would accept their prescriptions anyway.

Notes

1

REBUILDING AMERICA: EXPRESS HIGHWAYS AND VISIONS OF REFORM, 1890–1941

1. "The Magic City of Progress," *The American City* 54 (July, 1939), 41; *The New York Times*, April 6, 1939, May 17, 1939, May 19, 1939.
2. U.S. Department of Commerce, Bureau of Public Roads, *Highway Statistics: Summary to 1955* (Washington, D.C.: U.S. Government Printing Office, 1957), pp. 18–19, 25, 28; James J. Flink, *The Car Culture* (Cambridge: The MIT Press, 1975), pp. 18, 142, 147, 152–154; Robert S. Lynd and Helen M. Lynd, *Middletown: A Study in American Culture* (New York: Harcourt, Brace and World, 1929), pp. 258–259; U.S. Department of Commerce, Bureau of the Census, *Thirteenth Census of the United States, 1910*, Vol. 8 (Washington, D.C.: U.S. Government Printing Office, 1913), p. 45; U.S. Department of Commerce, Bureau of the Census, *Biennial Census of Manufactures*, 1925 (Washington, D.C.: U.S. Government Printing Office, 1928), p. 24.
3. Flink, *The Car Culture*, pp. 32–33, 35–36, 38; G. W. Atterbury, "The Commercial Car as a Necessity," *Harper's Weekly* 51 (December 28, 1907), 1925, as quoted in *ibid.*, p. 40; Glen E. Holt, "The Changing Perception of Urban Pathology: An Essay on the Development of Mass Transit in the United States," in Kenneth T. Jackson and Stanley K. Schultz, eds., *Cities in American History* (New York: Alfred A. Knopf, 1972), pp. 329–330, 332, 337; Henry James Carman, *The Street Surface Railway Franchises of New York City* (Columbia University Studies in History, Economics and Public Law, Vol. 88: New York, 1919), pp. 29–30, as quoted in *ibid.*, p. 329. See also James J. Flink, "Mass Automobility: An Urban Reform That Backfired," paper presented at the Missouri Valley History Conference, Omaha, Nebraska, March 6, 1975, pp. 3–4.
4. Flink, *The Car Culture*, p. 150; John C. Burnham, "The Gasoline Tax and the Automobile Revolution," *The Mississippi Valley Historical Review: A Journal of American History* 98 (December, 1961), 435, 442, 447–448, 456; BPR, *Highway Statistics: Summary to 1955*, pp. 12–13; National Highway Users Conference, *Dedication of Special Highway Revenues to Highway Purposes: An Analysis of the Desirability of Protecting Highway Revenues through Amendments to State Constitutions* (Washington, D.C.: National Highway Users Conference, 1941), p. 5. For additional evidence of road user support for the principle of gasoline taxation, see National

Highway Users Conference, *Highway Taxation, Finance and Administration: An Outline of Policies* (Washington, D.C.: National Highway Users Conference, 1938), pp. 7–9.

5. BPR, *Highway Statistics: Summary to 1955,* pp. 62–63, 68, 78; Flink, *The Car Culture,* p. 141; John B. Rae, *The Road and the Car in American Life* (Cambridge: The MIT Press, 1971), pp. 68, 71–72, 79–83; U.S. Congress, House, Committee on Roads, *Toll Roads and Free Roads,* House Document No. 272, 76th Congress, 1st Session, 1939, pp. 93, 95, 97, 104, 107–108, 110–111, 114, 197 (cited hereafter as *Toll Roads and Free Roads*); Henry A. Wallace, Secretary of Agriculture, to the President, February 13, 1939, OF 129, Franklin D. Roosevelt Library, Hyde Park, New York; Spencer Miller, Jr., head of the New Jersey Highway Department, "History of the Modern Highway in the United States," in Jean Labatut and William J. Lane, eds., *Highways in Our National Life: A Symposium* (Princeton, New Jersey: Princeton University Press, 1950), pp. 109–110; Richard O. Davies, author and editor, *The Age of Asphalt: The Automobile, the Freeway, and the Condition of Metropolitan America* (Philadelphia: J. B. Lippincott and Company, 1975), p. 3. In *Toll Roads and Free Roads,* MacDonald projected a 26,700-mile expressway network while the 1939 *Report* of the bureau called for a system no greater than 30,000 miles. One of MacDonald's deputies, Harold E. Hilts, calculated 29,330.7 miles; another, Herbert S. Fairbank, advertised a 28,000-mile system. Compare *Toll Roads and Free Roads,* p. 108; *Report of the Chief of the Bureau of Public Roads,* 1939, in Record Group 46, Senate, Records of the Committee on Post Offices and Post Roads, Doc. 132, 1940, National Archives, Washington, D.C. (cited hereafter as Senate, Records of the Committee on Post Offices and Post Roads); Herbert S. Fairbank, "Interregional Highways Indicated by State-Wide Highway Planning Surveys," *Roads and Streets* 83 (January, 1940), 37; Harold E. Hilts, "Planning the Interregional System," *Public Roads: A Journal of Highway Research* 22 (June, 1941), 94. For summaries of significant road-building plans and preliminary expressway development, see Roy Lubove, *Twentieth Century Pittsburgh: Government, Business, and Environmental Change* (New York: John Wiley and Sons, 1969), pp. 103–105; "A Prescription for Saving Downtown Cincinnati," *National Real Estate Journal* 41 (March, 1941), 16–18; Committee on Elevated Highways, American Road Builders' Association, "Report of Committee on Elevated Highways," *Proceedings of the Thirty-sixth Annual Convention of the American Road Builders' Association, 1939,* ed. Charles M. Upham (Washington, D.C.: American Road Builders' Association, 1939); pp. 293, 295; "Traffic Jams Business Out, Produces Bald Spots in City Centers," *Architectural Forum* 72 (January, 1940), 64–65 (Detroit); Harry W. Lochner, "Express Highways in the Chicago Metropolitan Area," Purdue University, *Proceedings of the Twenty-seventh Annual Road School,* Extension Series No. 50 (May, 1941), pp. 107–109.
6. Lloyd L. Sponholtz, "The Good Roads Movement in Ohio, 1900–1912," paper presented at the meeting of the Missouri Valley History Conference, March 12, 1976, pp. 7–13; *Toll Roads and Free Roads,* pp. 90, 93, 106–107; John B. Rae, "Coleman du Pont and His Road," *Delaware History* 16 (Spring-Summer, 1975), 171–172, 175–179; and Rae, *The Road and the Car*

in American Life, pp. 74, 76; Fred K. Stuart Green to Thomas H. MacDonald, February 15, 1938; clipping, *Washington Post,* March 3, 1938, both in Record Group 30, Bureau of Public Roads Classified Central File, 1912-1950, File 740.1.1 General 1938-37, Federal Records Center, Suitland, Maryland (cited hereafter as BPR Files). Even during the early days of auto and truck buying, road builders failed to keep in step with traffic increases. See James J. Flink, *America Adopts the Automobile, 1895-1910* (Cambridge, Massachusetts: The MIT Press, 1970), p. 213.

7. William H. Wilson, *The City Beautiful Movement in Kansas City* (Columbia, Missouri: University of Missouri Press, 1964), pp. 46, 48–53; *Kansas City Star,* December 6, 1893, as quoted in *ibid.,* p. 53.
8. Charles F. Puff, Jr., "Relation between the Small House and the Town Plan," *The Annals* 51 (January, 1914), 149–153; Charles M. Robinson, "Sociology of a Street Layout," *The Annals* 51 (January, 1914), 194. See also Jon A. Peterson, "The City Beautiful Movement: Forgotten Origins and Lost Meanings," *Journal of Urban History* 2 (August, 1976), 430.
9. Thomas Adams, *Planning the New York Region: An Outline of the Organization, Scope and Progress of the Regional Plan* (New York, 1927), pp. 28–29. For an account of road planning and civic boosterism in southern cities, see Blaine A. Brownell, "The Commercial-Civic Elite and City Planning in Atlanta, Memphis, and New Orleans in the 1920's," *The Journal of Southern History* 41 (August, 1975), 353–356; and his "The Automobile and Urban Planning in the 1920's: The Case of Three Cities in the American South," paper presented at the meeting of the Organization of American Historians, Denver, Colorado, April 19, 1974, pp. 8–10.
10. For an illustration of the variety of urban and social purposes road construction was supposed to serve, compare National Resources Committee, "The Division of Costs and Responsibilities for Public Works," October, 1935, pp. 14–15, in RG 187, Records of the National Resources Planning Board, File 732, National Archives, Washington, D.C. (cited hereafter as Records of the NRPB); Harold S. Buttenheim, "Urban Land Policies," in National Resources Committee, *Our Cities: Their Role in the National Economy* (Washington, D.C.: U.S. Government Printing Office, 1937), pp. 214, 236, 259, 271; Harland Bartholomew, head of one of the largest planning firms in the nation, "Effect of Urban Decentralization upon Transit Operation and Policies," *Proceedings of the American Transit Association and Its Affiliated Organizations, 1940* (New York: American Transit Association, n.d.), pp. 483–487; and Bartholomew's "The Neighborhood—Key to Urban Redemption," *American Planning and Civic Annual,* ed. Harlean James (Washington, D.C.: American Planning and Civic Association, 1941), p. 247; "A Prescription for Saving Downtown Cincinnati," pp. 16–18.
11. Flink, *The Car Culture,* pp. 7–8; Kenneth E. Peters, "The Good-Roads Movement and the Michigan State Highway Department, 1905–1917" (unpublished Ph.D. dissertation, The University of Michigan, 1972), pp. 34–35; and Peters, "Michigan Good-Roads Politics, 1900 to 1917," paper presented to the Missouri Valley History Conference, Omaha, Nebraska, March 12, 1976, p. 4; Sponholtz, "The Good Roads Movement in Ohio, 1900–1912," pp. 4–5; Rae, *The Road and the Car in American Life,* pp. 30–31; William E. Lind, "Thomas H. MacDonald: A Study of the Career of an Engineer

Administrator and His Influence on Public Roads in the United States, 1919–1953" (unpublished M.A. thesis, The American University, 1965), pp. 9–10; Frederic L. Paxson, "The American Highway Movement, 1916–1935," *American Historical Review* 101 (January, 1946), 238–239. See also Sam Bass Warner, Jr., *The Urban Wilderness: A History of the American City* (New York: Harper and Row, 1972), p. 37. For an account of the efforts of road engineers to secure highway improvements, see American Association of State Highway Officials, *American Association of State Highway Officials: A Story of the Beginning, Purposes, Growth, Activities and Achievements of AASHO* (Washington, D.C.: American Association of State Highway Officials, 1965), pp. 53, 152.

12. *Ibid.*, pp. 131, 151–152, 211–313; Peters, "Michigan Good-Roads Politics, 1900 to 1917," p. 3; Rae, *The Road and the Car in American Life*, pp. 37, 39, 74; 42 U.S. *Stat.*, 212; Paxson, "The American Highway Movement, 1916–1935," p. 247. See also Warner, *The Urban Wilderness*, p. 37; Charles L. Dearing, *American Highway Policy* (Washington, D.C.: The Brookings Institute, 1941), pp. 84–86, 173; Rae, "Coleman du Pont and His Road," p. 177.

    Gary T. Schwartz, "Urban Freeways and the Interstate System," *Southern California Law Review* 49 (March, 1976), 413–414, argues that state and federal officials ignored portions of the 1921 Act in order to concentrate funds on the primary network alone.

13. Henry F. Cabell, "The Economic Aspect of Interregional Highways," *Roads and Streets* 83 (January, 1940), 61. See also Murray D. Van Wagoner, Michigan State Highway Commissioner, "Superhighways Ahead," *Proceedings of the Thirty-sixth Annual Convention of the American Road Builders' Association, 1939*, pp. 16, 18.

14. P. H. Kitfield, "The Future of Highway Building in New England," *Proceedings of the Thirty-eighth Annual Convention of the American Road Builders' Association* (1941), ed. Charles M. Upham (Washington, D.C.: American Road Builders' Association, n.d.), p. 131. See also G. H. Delano, Chief Engineer, Massachusetts Department of Public Works, "Super-Highways and Primary Roads," *Proceedings of the Fourteenth Annual Convention of the Association of Highway Officials of the North Atlantic States* (Trenton, New Jersey: Office of the Secretary, 1938), p. 73; L. E. Boykin, Chief, Division of Highway Laws and Contracts, Public Roads Administration, "Interregional Highways: Legal and Right-of-Way Problems," *Roads and Streets* 83 (January, 1940), 57–59; Fred J. Grum, Engineer, California Highway Department, "California's Plan for Freeways in Metropolitan Areas," *Civil Engineering* 2 (October, 1941), 569.

15. Burnham, "The Gasoline Tax and the Automobile Revolution," p. 455; Committee on Elevated Highways, American Road Builders' Association, "Report of Committee on Elevated Highways" (1939), pp. 286, 288. See also Association of Highway Officials of the North Atlantic States, "Resolution," February, 1938, *Proceedings of the Fourteenth Annual Convention of the Association of Highway Officials of the North Atlantic States*, p. 259.

    In 1939, diversion amounted to 16 percent of state collected motor fuel taxes and 15 percent of all state user taxes (e.g., licenses, etc.). See U.S. Federal Works Agency, Public Roads Administration, *Highway Statistics:*

Summary to 1945 (Washington, D.C.: U.S. Government Printing Office, 1947), p. 37.

16. Chester H. Gray, *Transportation in 1950* (Washington, D.C.: National Highway Users Conference, c. 1940), p. 27; National Highway Users Conference, *The Eastman Report Finds that Highway Users Pay Their Way and More* (Washington, D.C.: National Highway Users Conference, 1940), p. 16; Roland Rice, "Roads Were Built for Commerce, Not Sightseeing," *Commercial Car Journal: The Magazine for Fleet Operators with Which Is Combined Operation and Maintenance* (November, 1938), 29. See also NHUC, *Dedication of Special Highway Revenues to Highway Purposes,* pp. 3, 5; NHUC, *Highway Taxation, Finance and Administration,* p. 7; Walter Mullady, Decatur Cartage Company and Vice President, Central Motor Freight Association, "Highway Haulage Not Subsidized," *Power Wagon: The Motor Truck Journal* 62 (March, 1939), 5–7. For road contractors' views of diversion, see "Resolutions Adopted by the Thirty-sixth Annual Convention of the American Road Builders' Association," March, 1939, in *Proceedings of the Thirty-sixth Annual Convention of the American Road Builders' Association,* p. 591.

    As early as 1921, military officers assigned to highway planning and liaison confirmed the supremacy of state and federal road engineers, arguing that highways adequate for commercial traffic were sufficient for military needs. See Stanley H. Ford, Lieutenant General, U.S. Army, "The Military Requirements of Our Highway System," *Proceedings of the Thirty-seventh Annual Convention of the American Road Builders' Association* (1940), ed. Charles M. Upham (Washington, D.C.: American Road Builders' Association, c. 1940), pp. 69–70; Dearing, *American Highway Policy,* pp. 138–139; A. W. Brandt, Superintendent of Public Works, New York State, "Shaping Our Highway Program for National Defense," *American Highways* (October, 1940), 18; Letter of Submittal from Henry A. Wallace, Secretary of Agriculture, concurred in by Harry H. Woodring, Secretary of War, to the President, April 11, 1939, in *Toll Roads and Free Roads,* p. x.

17. Arthur S. Link, et al., *American Epoch: A History of the United States since the 1890s,* Vol. 2 (New York: Alfred A. Knopf, 1967), p. 370; Milton Derber, "The New Deal and Labor," in John Braeman, et al., eds., *The New Deal: The National Level* (Columbus: Ohio State University Press, 1975), p. 123; Irving Bernstein, *The Lean Years: A History of the American Worker, 1920–1933* (Baltimore: Penguin Books, 1966), pp. 469–470; Albert U. Romasco, *The Poverty of Abundance: Hoover, the Nation, the Depression* (New York: Oxford University Press, 1968), p. 223; William E. Leuchtenburg, *Franklin D. Roosevelt and the New Deal, 1932–1940* (New York: Harper and Row, 1963), p. 121; BPR, *Highway Statistics: Summary to 1955,* p. 59. See also Rae, *The Road and the Car in American Life,* p. 74.

18. Richard Polenberg, "The Decline of the New Deal," in Braeman, et al., eds., *The New Deal: The National Level,* pp. 255–256; Daniel W. Bell to the President, May 14, 1938, OF 129; Franklin D. Roosevelt to the Acting Director of the Budget, May 16, 1938, OF IE, both in Roosevelt Library; Samuel I. Rosenman, ed., *Public Papers and Addresses of Franklin D. Roosevelt,* Vol. 8 (New York: The Macmillan Company, 1941), pp. 18, 229. See also James T. Patterson, *Congressional Conservatism and the New*

Deal: The Growth of the Conservative Coalition in Congress, 1933–1939 (Lexington: University Press of Kentucky, 1967), p. 238.

19. Memo for Files, March 17, 1939; Wilburn Cartwright, Chairman of the House Committee on Roads, to Franklin D. Roosevelt, May 15, 1939; Franklin D. Roosevelt to Edwin M. Watson, April 24, 1939, all in OF 129; Memo for Files, June 20, 1939, OF 3710, all in Roosevelt Library; "What Congress Plans to Do on Roads," *Engineering News-Record* 124 (January 18, 1940), 96; U.S. *Congressional Record*, 76th Congress, 3d Session, Vol. 86, Part 10, 11150. See also James Rowe, Jr., to Edwin M. Watson, April 24, 1939, OF 129, Roosevelt Library; Wallace to the President.

    As early as February, 1938, Roosevelt had asked MacDonald to secure details of excess condemnation practices in Great Britain. By the end of February, 1938, MacDonald was promoting the principle on his own initiative. See MacDonald to M. H. McIntyre, Secretary to the President, February 14, 1938, OF IE, Roosevelt Library; Thomas H. MacDonald to Harlean James, Executive Secretary of the American Planning and Civic Association, February 28, 1938, BPR File 740.1.1 General 1938-37, FRC. See also Franklin D. Roosevelt to M. H. McIntyre, February 16, 1938, OF IE, Roosevelt Library.

    MacDonald, to provide a counterbalance to Geddes' "bizarre ideas," sent Roosevelt a twelve-page digest of *Toll Roads and Free Roads,* a report on expressway and toll road development. See James Rowe, Jr., to Marguerite A. Lehand, March 22, 1939, OF 129, Roosevelt Library.

20. Franklin D. Roosevelt to the Secretary of the Interior and the Director of the Budget, April 1, 1939, OF 129, Roosevelt Library.
21. Lauchlin Currie to the President, June 21, 1940; John M. Carmody to the President, June 21, 1940; Franklin D. Roosevelt to Kenneth McKellar, June 21, 1940; Kenneth McKellar to the President, June 25, 1940, all in OF, IE; Franklin D. Roosevelt to the Administrator, Federal Works Agency, November 25, 1941, OF 129, all in Roosevelt Library; Harry S. Truman to Philip B. Fleming, Administrator, Federal Works Agency, September 6, 1945, OF 129, Harry S. Truman Library, Independence, Missouri.

    See also a public letter from Roosevelt to Kenneth McKellar and Wilburn Cartwright, November 21, 1941, OF 129, Roosevelt Library, in which the president urged them to cut spending on roads "not representing immediate requirements for our national defense."

22. *Toll Roads and Free Roads,* pp. 90, 94, 104.

2

PLANNING FOR POSTWAR AMERICA, 1941–1944

1. Mel Scott, *American City Planning since 1890: A History Commemorating the Fiftieth Anniversary of the American Institute of Planners* (Berkeley: University of California Press, 1969), p. 361; Rosenman, ed., *Public Papers and Addresses of Franklin D. Roosevelt,* Vol. 10, p. 13; Franklin D. Roose-

velt to Harland Bartholomew, April 14, 1941, OF 4388, Roosevelt Library; Pyke Johnson, "Highway Transportation in the Defense Effort," *Proceedings of the Thirty-eighth Annual Convention of the American Road Builders' Association*, p. 30. See also Wilburn Cartwright, "The Congressional Outlook," in *ibid.*, p. 32; Thomas H. MacDonald, "The New Federal Highway Program," *American Planning and Civic Annual*, 1941, p. 51; Herbert S. Fairbank, "Military Highways," University of Michigan, *Proceedings of the Twenty-seventh Annual Highway Conference* 43 (July, 1941), 43.

2. Public Law 295, 77th Congress, Section 9; Thomas H. MacDonald, "The City's Place in Post-War Highway Planning: Concrete Advice on What Cities Should Do to Help Provide the Economic 'Backlog' against Economic Recession," *The American City* 58 (February, 1943), 43. See also U.S. Congress, House, Committee on Roads, *Interregional Highways*, House Document No. 379, 78th Congress, 2d Session, 1944, p. 110 (cited hereafter as *Interregional Highways*).
3. Charles E. Merriam, "The National Resources Planning Board: A Chapter in American Planning Experience," *The American Political Science Review* 38 (December, 1944), 1075–1077, 1082.
4. Wilfred Owen, "Transportation and Public Promotional Policy," in [Advisory Committee for the Transportation Study], National Resources Planning Board, *Transportation and National Policy* (Washington, D.C.: U.S. Government Printing Office, 1942), pp. 267, 269, 276.
5. [Wilfred Owen], "The Future Development of Highway Transportation," Highway Research Board, *Proceedings of the Twenty-first Annual Meeting*, ed. Roy W. Crum (Washington, D.C.: Highway Research Board, c. 1942), p. 16; and his "Transportation and Public Promotional Policy," pp. 257, 259.
6. [Owen], "The Future Development of Highway Transportation," p. 16; Report of the American Society of Planning Officials on Highways and Transportation, as cited without further reference in Owen, "Transportation and Public Promotional Policy," p. 275; New York City Planning Commission, *Annual Report*, 1940, p. 38, as cited in *ibid.*; and Owen, "Highway Transportation: A Program to Meet the Impacts of War" (preliminary draft for technical review), November 16, 1942, Records of the NRPB, File 089, NA.
7. Wilfred Owen to C. M. Nelson, Editor, *Better Roads*, April 22, 1942, Records of the NRPB, File 732, NA; Owen, "Highway Transportation: A Program to Meet the Impacts of War," pp. 50–52; and his, "Transportation and Public Promotional Policy," pp. 275–276.
8. Letter of submittal from Owen D. Young to Frederic A. Delano, member of the NRPB, May 22, 1942; Letter of submittal from Frederic A. Delano, Charles E. Merriam, and George F. Yantis, members of the NRPB, to the President, May 25, 1942, both in [Advisory Committee for the Transportation Study], NRPB, *Transportation and National Policy*, pp. iii–v.

   See also Walter M. Blucher, Director, American Society of Planning Officials, "Planning for the Post-War Period," Purdue University, *Proceedings of the Twenty-eighth Annual Road School, January, 1942*, Vol. 26, No. 2 (Lafayette: Purdue University, 1942), 31, 33–34.

9. Memo for Files, March 31, 1941, OF IE; James Rowe, Jr., to the President, April 11, 1941; Public letter from Franklin D. Roosevelt to John M. Carmody, April 14, 1941, both in OF 4388, all in Roosevelt Library.

   In effect, only Bartholomew, Kennedy, Purcell, and MacDonald, with his own assistants, attended most committee meetings and wrote the final report. Between April, 1941, and early 1944, when their report was finished, Delano withdrew from an active role, Graves died, and Tugwell accepted appointment as governor of Puerto Rico. See Frederic A. Delano to the President, April 19, 1941, Records of the NRPB, File 732, NA; Letter of Submittal from National Interregional Highway Committee to Philip B. Fleming, January 1, 1944, *Interregional Highways,* p. x.

   On July 22, 1941, Carmody asked Roosevelt to appoint a military man to the committee in order to secure his opinion on routing. On August 29, Harold D. Smith, Director, Bureau of the Budget, wrote to Roosevelt that "necessary information could be secured through consultations with representatives of the War Department or any other department." That day, Roosevelt told Carmody to "secure the advice of departments through proper liaison." See Harold D. Smith to the President, August 29, 1941, OF 4388, Roosevelt Library; Franklin D. Roosevelt to John M. Carmody, August 29, 1941, in National Interregional Highway Committee, Minutes of the Second Meeting, September 8, 1941, Records of the NRPB, File 732, NA.
10. *Ibid.*
11. *Interregional Highways,* pp. 20–21, 61, 64–65, 126.
12. *Ibid.,* pp. 53–55, 64, 70, 87–88. By 1944, members of the Interregional Committee had dropped unemployment as a guide to route location. Road building was necessary and not contingent on unemployment, they claimed, but there was a "remarkable correlation" between location of proposed Interregional highways and areas of projected joblessness. See *ibid.,* p. 32. MacDonald was particularly adamant about this point. See Notes of Thomas H. MacDonald's remarks before the meeting of the Western Association, San Francisco, July 4, 1943, p. 7, File 790 California 1949-43; and Thomas H. MacDonald to Charles H. Purcell, January 30, 1943, 481 California FAS, both in BPR Files, FRC.
13. Cincinnati. The City Planning Commission, *The Official City Plan of Cincinnati, Ohio* (Cincinnati: The City Planning Commission, 1925), pp. 48–54, 87–91, 227–229; Tracy B. Augur to Herbert S. Fairbank, September 25, 1944, BPR File 481 Corresp. 1943-1944, FRC; Scott, *American City Planning since 1890,* pp. 437–440.
14. E. E. East, Chief Engineer, Automobile Club of Southern California, to Frank W. Herring, Assistant Director of the NRPB, January 23, 1942, Records of the NRPB, File 732, NA; E. E. East to L. I. Hewes, Chief, Western Region, PRA, May 6, 1942; E. E. East to Thomas H. MacDonald, May 11, 1942; Memorandum for File from Charles C. Morris, District Engineer, PRA, March 16, 1944, all in BPR File 740.1.2 California 1945–1938, FRC.
15. Arthur S. Dudley, "Sacramento's Post-War Plans Two Major Fronts: Public Works and Private Industry," *The American City* 58, Part 2 (October, 1943), 45; District Engineer, T-1, to Toledo Chamber of Commerce, July 15, 1943; John D. Carney, Postwar Planning, City Engineering Department,

to Thomas H. MacDonald, February 8, 1944, both in BPR File 481 Corresp. FAS Ohio 1943-44, FRC.

16. For Highway Department plans, see Hal G. Sours to C. E. Swain, April 6, 1943; Report by I. W. Hall on FAPS (Federal Aid Primary System) Rt. No. 152, April 14, 1948; Frank B. Johnson to Harold E. Hilts, June 22, 1944; Thomas H. MacDonald to Harry L. Linch, December 14, 1944, all in BPR File 481 Ohio FAS, FRC.
17. Arthur F. Hewitt, "Freeways," in University of Colorado, *Proceedings of the Colorado Highway Conference* (Boulder, n.p., 1942), p. 21; Charles H. Purcell to Lawrence I. Hewes, January 14, 1943, File 481 California FAS; George T. McCoy, by C. R. Montgomery, to L. E. Boykin, May 4, 1943, File 740.1.1 General 1943; Ohio State Highway Department, *Discussion of Preliminary Engineering Reports for Advanced Review,* August, 1944, pp. 13–14, File 481 Ohio FAS, all in BPR Files, FRC; R. C. Chaney, "Cleveland Freeways a Major Problem," *Roads and Streets* 86 (June, 1943), 42–43; Roy E. Jorgensen, "Report of the Bureau of Highway Planning Studies," State of Connecticut, *Biennial Report of the Highway Commissioner to the Governor for the Fiscal Years Ended June 30, 1943, and June 30, 1944,* Public Document No. 36 (published by the state, 1944), p. 91. See also State of Michigan, *Twentieth Biennial Report of the State Highway Commissioner for the Fiscal Years Ending June 30, 1943, and June 30, 1944* (By Authority, n.d.), p. 29; David R. Levin, Public Roads Administration, "Limited Access Highways in Urban Areas," *The American City* 59 (February, 1944), 77.
18. Sidney J. Williams, "Safety in War and Postwar Projects," an address before the Mississippi Valley Conference of State Highway Officials, January 29, 1944, in BPR File 016 (Gen.) 1944-45, FRC.
19. Remarks of J. Will Robinson, Chairman, House Committee on Roads, February 29, 1944, in U.S. Congress, House, Committee on Roads, *Hearings on H.R. 2426: Federal Aid for Post-War Highway Construction,* Vol. 1, 78th Congress, 2d Session, 1944, p. 5.
20. H.R. 2426, the AASHO bill, was published in *ibid.*, pp. 1–3.
21. Resolution of Chambers of Commerce of Salina and Pratt, Kansas, circa January, 1944; H. B. Henderlite, Louisiana Department of Highways, to Governor Sam H. Jones, January 25, 1944; Tom Stewart to H. E. Sargent, Commissioner, Vermont Highway Department, December 17, 1943, all in BPR File 740.1.1 1944, FRC.

    See also Thomas D. Winter, member of Congress from the Third District, Kansas, to Dear Colleague, June 14, 1944; J. A. Elliot, PRA District Engineer, to Thomas H. MacDonald, September 25, 1944, both in *ibid.*; Resolution from the Directors of the Southeastern Association of State Highway Officials to the Resolutions Committee, AASHO, meeting in Annual Convention, November 27, 1944, BPR File 016 (Gen.) 1944-45, FRC.
22. Letter from Brady P. Gentry, President of AASHO, October 14, 1943 (recipient not clear), File 740.1.1 1943; William J. Cox to Robert Moses, December 14, 1943; William J. Cox to Herbert S. Fairbank, May 17, 1944; Spencer Miller, Jr., New Jersey Highway Commissioner, to Thomas H. MacDonald, May 23, 1944; Telegram from William J. Cox, et al., to Franklin D. Roosevelt, April 14, 1944, all in BPR File 740.1.1 1944; *Washington*

Post, December 19, 1943; A. Lee Grover, Secretary-Treasurer, Association of Highway Officials of the North Atlantic States, to Kenneth McKellar, March 1, 1943, Senate, Records of the Committee on Post Offices and Post Roads, File 78A-F23, NA; H.R. 4628, the bill prepared by dissident road officials and remarks of Spencer Miller, Jr., in Committee on Roads, House, *Hearings on H.R. 2426: Federal Aid for Postwar Highway Construction,* pp. 1049–1055.

See also S. C. Hadden, President of AASHO, December 21, 28, 29, 1943, to Thomas H. MacDonald, File 016 General 1944-45; Robert Moses to G. Donald Kennedy, Vice President, Automotive Safety Foundation, March 4, 1944, File 740.1.1 1944, both in BPR Files; Edward J. Kelley, Mayor of Chicago, to Kenneth McKellar, June 24, 1943, Senate, Records of the Committee on Post Offices and Post Roads, File 78A-F23, NA. Remarks of Fiorella LaGuardia, Mayor of New York, in *Proceedings of the Twentieth Annual Meeting of the Association of Highway Officials of the North Atlantic States* (Trenton, New Jersey: The Association of Highway Officials of the North Atlantic States, Office of the Secretary, State House Annex, 1944), p. 7.

23. "State Highway Departments Urged to Speed Plans for Post-War Work," *Highway Highlights: Automotive Transportation in All Its Phases* (Washington, D.C.: National Highway Users Conference, October 20, 1943), p. 1; Russell E. MacCleery, "The Place of the Motor Vehicle in Post-War Transportation," *Proceedings of the Twentieth Annual Meeting of the Association of Highway Officials of the North Atlantic States,* pp. 65, 71; Texas Good Roads Association, *Highway Highlights,* July 15, 1944, located in BPR File 740.1.1 1944, FRC. See also A. B. Barber, Transportation and Communication Department, U.S. Chamber of Commerce, to Thomas H. MacDonald, April 13, 1944, in *ibid.*

24. Hal H. Hale to Members of the Executive Committee, AASHO, August 19, 1944, BPR File 016 General 1944-45, FRC; Text of S. 2105 in U.S. Congress, Senate, Committee on Post Offices and Post Roads, *Post-War Federal-Aid Highway Act of 1944,* 78th Congress, 2d Session, 1944, Sen. Report No. 1056 to accompany S. 2105, pp. 7–9.

    Hayden himself had long been active in highway development and road legislation, sponsoring road bills, cosponsoring antidiversion legislation, and fighting toll road schemes. During the war, he stayed in touch with Commissioner MacDonald about road legislation and requested a report on postwar highway development from the Automotive Safety Foundation, facts which together suggest a high level of personal interest in highway matters.

25. U.S. *Congressional Record,* 78th Congress, 2d Session, 1944, Vol. 90, Part 6, 7684, 7784–7785, 7787–7791, 7795, 7797–7798, 7806–7807, 8506, 8509–8511, 8530. By a vote of 47–8, members of the Senate turned down a motion by Arthur H. Vandenberg to reconsider Russell's amendment.

26. *Ibid.,* Part 7, 9272–9275.

27. *Ibid.,* Part 6, 7676.

28. Philip B. Fleming to the President, January 5, 1944, OF 4388, Roosevelt Library; Message from the President to the Congress of the United States, January 12, 1944, in *Interregional Highways,* p. iv. See also Harold D.

Smith, Director, Bureau of the Budget, to the President, March 2, 1944; Franklin D. Roosevelt to Philip B. Fleming, March 3, 1944; Franklin D. Roosevelt to William J. Cox, April 15, 1944, all in OF 129, Roosevelt Library.

3

THE POLITICS OF HIGHWAY FINANCE, 1945–1950

1. Randall R. Howard, "Coast-to-Coast Trucking Shows Steady Upward Trend," *Power Wagon: The Motor Truck Journal* 83 (December, 1949), 9–11; Link, et al., *American Epoch: A History of the United States Since the 1890's*, Vol. 3, pp. 606–608; Fred B. Lautzenheiser, "Effect of the AASHO Code on Truck Design," *Commercial Car Journal: The Magazine for Fleet Operators with Which Is Combined Operation and Maintenance* (July, 1946), 38–39.
2. "Long Distance Trucking Handicapped by Highway Shortcomings," *Power Wagon: The Motor Truck Journal* 77 (July, 1946), 9; P. Hirsch, "Chicago Fleetmen Battle Traffic Bottlenecks," *Commercial Car Journal: The Magazine for Fleet Operators with Which Is Combined Operation and Maintenance* (June, 1954), 68, 70; Transportation and Communication Department, United States Chamber of Commerce, "Merchandise Pickup and Delivery," *Power Wagon: The Motor Truck Journal* 85 (August, 1950), 9.
3. Public Roads Administration, "Explanation and Interpretation of the Federal Aid Highway Act of 1944," April, 1945, pp. 4, 10–11, BPR File 740.1.1 General 1939-1950, FRC. See also "Federal Aid Highway Act of 1944" (Draft), January 5, 1945, File 740.1.1 General 1945-1944; Lawrence S. Tuttle, Public Roads Administration, to the Commissioner (of Public Roads), June 6, 1945, File 740.1.1 General 1939-1950, both in BPR Files, FRC.

   For efforts to secure special routings for army maneuvers and for commercial and agricultural development, see Report of Conference between B. J. Welch, Thomas H. MacDonald, et al., February 21, 1945; Floyd P. Willette, Executive Secretary, Council Bluffs Chamber of Commerce, to Representative Ben F. Jensen, March 13, 1945, both in File 740.1.1 1946-45 General; Harry L. Benbough, Jr., President, El Cajon Valley Chamber of Commerce, to Thomas H. MacDonald, October 24, 1946; Jack Hanna, President, La Mesa Chamber of Commerce, to Thomas H. MacDonald, October 26, 1946; George H. Robison, Secretary, El Cajon Boulevard Civic Association (San Diego), to Thomas H. MacDonald, October 28, 1946, all in File 481 Corresp. FAS Calif. 1946, all in BPR Files; Memo prepared by General C. P. Gross, Chief of Transportation, Department of the Army, October 19, 1945; "Indicated Changes and Additions in the 'Interregional Highway System,'" November, 1945, pp. 1–2; "Procedure for Designating Defense Highways," November, 1945; General Paul F. Yount, Assistant Chief of Transportation, to the Chief of Transportation, October 21, 1947; Colonel Ross B. Warren, Office of the Chief of Transportation, to Herbert

S. Fairbank, December 29, 1948; Major George N. Davies, Chief, Highway Transport Branch, "Highways Essential for National Defense Within the Second Army Area," 1949, all in Department of the Army, Records of the Office of the Chief of Transportation, File 611, Federal Records Center, Suitland, Maryland. See also Chester Wardlow, *United States Army in World War II, the Technical Services, the Transportation Corps: Responsibilities, Organization, and Operations* (Washington, D.C.: United States Army, Office of the Chief of Military History, 1951), pp. 366–367.

4. S. F. Hoffman, District Engineer, to the Commissioner (of Public Roads), June 2, 1945; "Routes Proposed for Inclusion in the National System of Interstate Highways: Auxiliary Circumferential and Distributing Highways," both in BPR File 481 Corresp. FAS Ohio 1945, FRC; Schwartz, "Urban Freeways and the Interstate System," p. 424; AASHO, *The American Association of State Highway Officials,* p. 184. See also "Texas Urban Expressways Being Designed from Center Out," *Roads and Streets* 88 (November, 1945), 65–68.
5. BPR, *Highway Statistics: Summary to 1955,* p. 28.
6. *Ibid.,* pp. 18, 22–23; National Highway Users Conference, *Reports,* December 22, 1949, p. 2, Papers of Lou E. Holland, Truman Library.
7. U.S. Federal Works Agency, Public Roads Administration, *Work of the Public Roads Administration, Annual Report for the Fiscal Year 1948* (Washington, D.C.: U.S. Government Printing Office, n.d.), p. 1; U.S. Federal Works Agency, Public Roads Administration, *Work of the Public Roads Administration, Annual Report for the Fiscal Year 1949* (Washington, D.C.: U.S. Government Printing Office, 1950), p. 1; U.S. Bureau of Public Roads, *Annual Report: Fiscal Year 1952* (Washington, D.C.: U.S. Government Printing Office, n.d.), p. 2; U.S. Department of Commerce, *Highway Statistics: 1955* (Washington, D.C.: U.S. Government Printing Office, 1957), pp. 166–167; U.S. Department of Transportation, *Highway Statistics: Summary to 1965* (Washington, D.C.: U.S. Government Printing Office, 1967), pp. 65–66, 74–75; Roy E. Jorgensen, "Financing the Highway Program," *Proceedings of the Thirty-fourth Annual Convention of the American Association of State Highway Officials* (Washington, D.C.: American Association of State Highway Officials, 1948), p. 58; Thomas H. MacDonald, "A Review of Highway Developments," February 12, 1953, pp. 3–4, paper to the AAA Highway Emergency Conference, Holland Papers, Truman Library.
8. U.S. Bureau of Public Roads, *Annual Report: Fiscal Year 1951* (Washington, D.C.: U.S. Government Printing Office, n.d.), pp. 66–67; BPR, *Annual Report: Fiscal Year 1952,* p. 4; D. W. Ormsbee, Engineer, Colorado Highway Department, "The North-South Interstate Highway in Colorado," *The Twenty-third Annual Highway Conference of the University of Colorado, 1950* (Boulder: The University of Colorado, Highway Series, 1950), p. 61; C. M. McCormack, Consulting Engineer, Automotive Safety Foundation, "Ohio's Highway Needs," *Proceedings of the Ohio Highway Engineering Conference, 1951* (Columbus: The Ohio State University, 1951), p. 44; Notes for American Association of State Highway Officials–Associated General Contractors of America, Incorporated, Joint Cooperative Committee Meeting, San Francisco, California, February 28, 1950, prepared by Richard

Wilson, California Division of Highways, in Papers of the American Association of State Highway Officials–Associated General Contractors of America, Incorporated, 1951, Washington, D.C.: Associated General Contractors of America, Incorporated.

9. National Highway Users Conference, *Information Service,* December 29, 1949, pp. 2–4; NHUC, *With State Highway Users Conferences,* January 3, 1950; NHUC, *Reports,* June 9, 1950, June 21, 1950, all in Holland Papers, Truman Library. See also Arthur C. Butler, "Highways Are Your Business," June 11, 1951, p. 8 (address before the New England Motor Carriers Conference), Holland Papers, Truman Library.
10. Arthur C. Butler, "Road Blocks to Highway Progress," Purdue University, *Proceedings of the Thirty-fifth Annual Road School* 33 (Purdue University: Lafayette, 1949), 43; Butler, "Legislative Outlook for 1949," *Commercial Car Journal: The Magazine for Fleet Operators with Which Is Combined Operation and Maintenance* (November, 1948), 98–99; NHUC, *Texts of Good Roads Amendments: State Constitutional Provisions Safeguarding Highway User Taxes* (Washington, D.C.: National Highway Users Conference, 1949), pp. 8–15, 18–28, available in Holland Papers, Truman Library; BPR, *Highway Statistics: Summary to 1955,* p. 48.
11. C. R. Montgomery, "Analysis of Collier-Burns Highway Act of 1947," June 26, 1947, in BPR File 740.1.2 California 1950-46, FRC.
12. BPR, *Highway Statistics: Summary to 1955,* p. 74; M. E. Cox, "A Highway and Transportation Plan Emerges," *Civil Engineering: The Magazine of Engineered Construction* 24 (March, 1954), 143.
13. Lawrence I. Hewes to Thomas H. MacDonald, November 30, 1948, BPR File 481 Calif. FAS General (Acc. No. 58-A-778), FRC; G. P. St. Clair, "Bond-Issue Financing of Arterial Highway Improvements," Highway Research Board, *Proceedings of the Twenty-ninth Annual Meeting, 1949* (Washington, D.C.: Highway Research Board, 1950), p. 43; Remarks of Herbert S. Fairbank to the Second Highway Transportation Congress, May 6, 1948, *Proceedings of the Second Highway Transportation Congress* (Washington, D.C.: National Highway Users Conference, 1948), p. 20; Herbert S. Fairbank to Thomas H. MacDonald, November 30, 1948, BPR File 740.1.1 General 1949-48, FRC.
14. "A Petition to the Congress"; Arthur C. Butler to Harold Knutson, Chairman, Ways and Means Committee, July 15, 1947; List of signatories to Petition to the Congress, all published in *Highway Highlights: Automotive Transportation in All Its Phases* (August, 1947), 5–7; Summary of Minutes of Joint Meeting, Board of Governors and Administrative Committee, National Highway Users Conference, December 14, 1949, Holland Papers, Truman Library. See also *Highway Highlights: Automotive Transportation in All Its Phases* (January-February, 1950), 1, 5, 7; National Highway Users Conference pamphlet, published November, 1949, pp. 1, 4, 6, 9, 11; NHUC, *Information Service,* January 25, 1950, p. 2; Arthur C. Butler to A. J. Montgomery, American Automobile Association, January 26, 1950, all in Holland Papers, Truman Library; Remarks of C. H. Buckius, Assistant Chief Engineer, Pennsylvania Department of Highways, to the Third Highway Transportation Congress, April, 1950, *Proceedings of the Third High-*

way Transportation Congress (Washington, D.C.: National Highway Users Conference, 1950), pp. 35–36.

15. Butler to Knutson; Summary of Minutes of Joint Meeting, Board of Governors and Administrative Committee, National Highway Users Conference, December 14, 1949; Dawes E. Brisbane, Research Counsel, NHUC, "What Percentage of Highway Construction Costs Should Be Paid by the Federal Government," November 8, 1949, pp. 2–3 (panel discussion before Fall Conference of the American Road Builders' Association), Holland Papers, Truman Library; Remarks of Russell E. Singer to the Third Highway Transportation Congress, April, 1950, *Proceedings of the Third Highway Transportation Congress,* pp. 36–37.
16. Summary of Minutes of Joint Meeting, Board of Governors and Administrative Committee, National Highway Users Conference, December 14, 1949.
17. "Work of the President's Highway Safety Conference," c. 1949, Kenneth Hechler Files; Harry S. Truman to Philip B. Fleming, October 12, 1946, OF 140; "Economic Report of the President," (Confidential) materials presented by Council of Economic Advisers in personal meeting with the President, December 31, 1947, p. 75; Papers of John D. Clark, all in Truman Library; *Highway Highlights: Automotive Transportation in All Its Phases* (August, 1946), 4.
18. Robinson Newcomb to Spencer Miller, Jr., February 12, 1948, BPR File 740.1.1 General 1948, FRC; "Economic Report of the President," materials presented by Council of Economic Advisers in personal meeting with the President, December 31, 1947, p. 75; James E. Webb to the President, January 30, 1948, OF 129, Truman Library.

    Between 1946 and 1948, federal officials collected $1.3 billion from motor fuel taxes and $1.68 billion from taxes on vehicle and parts sales. They spent $812 million to build roads, including those constructed by the Bureau of Land Management and Corps of Engineers. But MacDonald administered construction of highways of greatest interest to road users and state highway engineers, and he spent $795 million, or about $505 million less than the treasury collected from gasoline tax revenues alone. See Department of Transportation, *Highway Statistics: Summary to 1965,* pp. 52–53, 169.
19. Remarks of H. Willis Tobler, American Farm Bureau Federation, to the Second Highway Transportation Congress, May 6, 1948, *Proceedings of the Second Highway Transportation Congress,* p. 19; Tobler, "The Farmers Need for Good Roads," *ibid.,* p. 16; Albert S. Goss, Master, The National Grange, to J. Harry McGregor, March 26, 1948, BPR File 740.1.1 General 1948, FRC; R. Flake Shaw, North Carolina Farm Bureau, to Senator John J. McClellan, May 17, 1949, published in U.S. Congress, Senate, Subcommittee of the Committee on Public Works, *Hearings on S. 244 and S. 1471, Rural Local Roads,* 81st Congress, 1st Session, 1949, p. 173. See also letters and testimony in *ibid.*
20. Milton R. Young to H. Alexander Smith, June 10, 1949, BPR File 740.1.1 General 1950-49, FRC; bill published in *Hearings on S. 244 and S. 1471, Rural Local Roads,* pp. 1–3.
21. Spencer Miller, Jr., to John C. Stennis, April 20, 1949; Hal H. Hale to George W. Malone, May 24, 1949, both in BPR File 740.1.1 General 1949-

48, FRC. See also Arthur R. Siegle, Public Roads Administration, to the Commissioner, March 4, 1949; F. R. White, Iowa State Highway Commissioner, to the Commissioner, May 28, 1949, both in BPR File 740.1.1 General 1949-48; G. P. St. Clair to Roy E. Jorgensen, August 25, 1949, BPR File 740.1.1 General 1950-49, all at FRC; Statement of Thomas H. MacDonald and statement of Carl W. Brown to the Senate Subcommittee on Roads, in *Hearings on S. 244 and S. 1471, Rural Local Roads,* pp. 26, 317.

22. A Bill Proposed by AASHO and sent to Senator Dennis Chavez, from a Special Meeting, AASHO, Chicago, November 21, 1949, Records of the Senate Public Works Committee, File 81A-F14, NA; Policy Statement of the American Association of State Highway Officials, adopted November 28, 1949, published in U.S. Congress, House, Committee on Public Works, *Hearings on H.R. 7398 and H.R. 7941, Federal-Aid Highway Act of 1950,* 81st Congress, 2d Session, 1950, pp. 224–225; Draft of Legislation in *ibid.,* pp. 225–226, 228–229.

23. Brisbane, "What Percentage of Highway Construction Costs Should Be Paid by the Federal Government," p. 6; Harold F. Hammond, "What Percent of Highway Construction Costs Should Be Paid by the Federal Government," *Traffic Engineering* 20 (March, 1950), 230; Samuel C. Hadden, Indiana Highway Department, to Herbert S. Fairbank, July 1, 1950, BPR File 740.1.1 General 1950, FRC.

24. House, Committee on Public Works, *Hearings on H.R. 7898 and H.R. 7941, Federal-Aid Highway Act of 1950,* pp. 1–2, 4; U.S. Congress, House, Committee on Public Works, *Amending and Supplementing the Federal-Aid Road Act,* 81st Congress, 2d Session, 1950, H. Rept. 2044 to accompany H.R. 7941, pp. 10–13.

25. U.S. *Congressional Record,* 81st Congress, 2d Session, 1950, 96, Pt. 6, pp. 7336, 7338, 7344–7345, 7349.

    Keating's motion to recommit was defeated 113–24 and 140–18; McGregor's motion lost 20–34 and 17–70. The committee bill passed 246–34, with 29 of the 34 dissenting votes cast by representatives from Massachusetts, Michigan, New Jersey, New York, and Pennsylvania.

26. U.S. Congress, Senate, Committee on Public Works, *Amending and Supplementing the Federal-Aid Road Act,* 81st Congress, 2d Session, 1950, S. Rept. No. 2044, to accompany H.R. 7941, pp. 3, 5, 7, 9.

27. *Congressional Record,* 96, Pt. 9, pp. 12692, 12704–12705, 12788, Pt. 10, pp. 12980–12982, 12984.

    States west of the Mississippi and south of the Ohio, Chavez warned his colleagues, would be "worse off" under the provisions of the Lodge amendment. Actually, Florida, Louisiana, Virginia, and West Virginia would have gained small amounts under a population formula. See *ibid.,* Pt. 9, pp. 12726, 12728.

28. Harry S. Truman to Dennis Chavez, August 17, 1950, OF 129, Truman Library.

29. *Congressional Record,* 96, Pt. 9, p. 12790, Pt. 10, pp. 12974, 12984–12986, 12992–12993, 13006. See also National Highway Users Conference, *Information Service,* August 29, 1950, Holland Papers, Truman Library.

30. *Congressional Record,* 96, Pt. 10, pp. 13706–13707, 13713, 13716.

31. National Highway Users Conference, *Information Service*, May 18, 1950, Holland Papers, Truman Library.

4

PROJECT ADEQUATE ROADS: TRAFFIC JAMS, BUSINESS, AND GOVERNMENT, 1951–1954

1. Arthur C. Butler, "Our Highway Problem and Some Recommendations for Meeting It," presented to the Joint Meeting of the Board of Governors and Administrative Committee of the National Highway Users Conference, October 11, 1951; Summary of Minutes of Joint Meeting, Board of Governors and Administrative Committee, National Highway Users Conference, October 11, 1951, both in Holland Papers, Truman Library; Transportation and Communication Department, U.S. Chamber of Commerce, "Merchandise Pickup and Delivery," p. 9; Hirsch, "Chicago Fleetmen Battle Traffic Bottlenecks," pp. 70–71, 180, 182.
2. Butler, "Our Highway Problem and Some Recommendations for Meeting It"; impressions gained from study of trucking industry publications. See also American Automobile Association, Highway Committee (presented by William A. Stinchcomb), "The Highway Situation in the National Emergency," presented at the forty-ninth annual meeting of the American Automobile Association, October 23, 1951; Hal H. Hale, Executive Secretary of AASHO, "Toll Roads," presented to the AAA Highway Emergency Conference, October 12, 1953, Washington, D.C., both in Holland Papers, Truman Library.
3. Butler, "Our Highway Problem and Some Recommendations for Meeting It."
4. "ATA Convention Pinpoints Highway Problems," *Commercial Car Journal: The Magazine for Fleet Operators with Which Is Combined Operation and Maintenance* (December, 1951), 64; Arthur M. Hill, Vice-Chairman, NHUC, to Lou E. Holland, President, American Automobile Association, October 24, 1951, Holland Papers, Truman Library; Butler, "Our Highway Problem and Some Recommendations for Meeting It"; Summary of Minutes of Joint Meeting, Board of Governors and Administrative Committee, NHUC, October 11, 1951.
5. Butler, "Our Highway Problem and Some Recommendations for Meeting It."
6. Roy E. Jorgensen, remarks to a Joint Meeting of the Board of Governors and Administrative Committee of the NHUC, in *ibid.*; "ATA Convention Pinpoints Highway Problems," pp. 64–66.

   Reports of enthusiastic support for highway needs studies and sufficiency rating methods before the founding of PAR in R. H. Baldock, "The Highway Situation Today and What Oregon Is Doing about It"; remarks of Walter Graf to the AAA Round-Table Discussion; Pyke Johnson and Carl E. Fritts, "State Highway Planning Studies," all presented at AAA Highway Emergency Conference, February 12–13, 1953; National Highway

Users Conference, *With State Highway Users Conferences*, February 20, 1951, all in Holland Papers, Truman Library; [Automotive Safety Foundation], *An Engineering Study of Ohio's Highways, Roads and Streets: A Report to the Ohio Program Commission and the Highway Study Committee* (n.p., n.d.); Butler, "Legislative Outlook for 1949," pp. 99–101, 182; Fred Harter, "The Ohio Highway Study Committee, Its Function and Purpose," *Proceedings of the Ohio Highway Engineering Conference, 1950* (Columbus: The Ohio State University, 1950), pp. 38–39. See also William E. Willey, "Arizona Highway Sufficiency Rating System," *Proceedings of the Thirty-fourth Annual Convention of the American Association of State Highway Officials* (Washington, D.C.: American Association of State Highway Officials, 1948), pp. 30–37; Willey, "Measurement of Highway Needs by Sufficiency Ratings," *Proceedings of the Thirty-seventh Annual Convention of the American Association of State Highway Officials* (Washington, D.C.: American Association of State Highway Officials, 1951), pp. 22–24.

7. Jorgensen, remarks to a Joint Meeting of the Board of Governors and Administrative Committee, NHUC.
8. "ATA Convention Pinpoints Highway Problems," p. 65; Roy E. Jorgensen, "Better Roads with PAR," *Proceedings of the Ohio Highway Conference, 1952* (Columbus: The Ohio State University, 1952), p. 67; Arthur C. Butler, "PAR—What It Is and What It Has Done," remarks to AAA Highway Emergency Conference, February 12, 1953, Holland Papers, Truman Library; Butler, "Our Highway Problem and Some Recommendations for Meeting It."
9. Henry K. Evans, "Can We Afford Model T. Roads?" *Commercial Car Journal: The Magazine for Fleet Operators with Which Is Combined Operation and Maintenance* (August, 1952), 120; Butler, "PAR—What It Is and What It Has Done"; National Highway Users Conference, *With State Highway Users Conferences*, December 12, 1952, Holland Papers, Truman Library.
10. Jorgensen, remarks to a Joint Meeting of the Board of Governors and Administrative Committee of the NHUC; and "Better Roads with PAR," p. 67; and "Sizes of State Highway Systems: How Big Should They Be?" *Traffic Quarterly* 6 (January, 1952), 66; and, *An Analysis of the Highway Program* (n.p., NHUC, 1951), pp. 15–17.
11. Bulletin from AASHO-AGC Joint Cooperative Committee to Secretaries, Managers and Presidents of AGC Chapters, Members of Joint Cooperative Committee, AASHO-AGC, and Officials of State Highway Departments and Bureau of Public Roads, December 18, 1950, March 20, 1951, both in Papers of the American Association of State Highway Officials–Associated General Contractors of America, Incorporated, Joint Cooperative Committee; Henry K. Evans, "The Great Highway Robbery—Or Is It?" *Commercial Car Journal: The Magazine for Fleet Operators with Which Is Combined Operation and Maintenance* (September, 1953), 110. The President of AASHO, Charles M. Ziegler, spoke of a fifteen-year program. See his remarks in U.S. Congress, House, Committee on Public Works, *National Highway Study*, Pt. 1, 83rd Congress, 1st Session, 1953, p. 203.
12. Report on S. 2437 from Frank Pace, Jr., to Dennis Chavez, February 19, 1952, Records of the Senate Public Works Committee, 82nd Congress, File

2437, Doc. 78, NA; Frank Pace, Jr., to Frederick W. Lawton, June 17, 1952; Roger W. Jones to William J. Hopkins, June 24, 1952, both in White House Bill File, H.R. 7340, Truman Library; DeWitt C. Greer, "Balancing the Highway Needs for Both Rural and Urban Areas," *Traffic Quarterly* 6 (July, 1952), 333–334; William A. Bresnahan, "Who Should Pay How Much of Highway Costs?" *Commercial Car Journal: The Magazine for Fleet Operators with Which Is Combined Operation and Maintenance* (July, 1952), p. 74. See also American Trucking Associations, *Statement of Highway Policy* (Washington, D.C.: American Trucking Associations, 1951), p. 4; William A. Bresnahan, "Truck Transportation—From the Truckers Viewpoint," *Proceedings of the Ohio Highway Engineering Conference, 1951*, pp. 11–12.

13. Charles Sawyer, Secretary of Commerce, to the President, June 18, 1952, White House Bill File, H.R. 7340; Jones to Hopkins; 66 U.S. *Stat.*, 158–159.
14. Bresnahan, "Who Should Pay How Much of Highway Costs?" p. 244; see also Butler, "PAR—What It Is and What It Has Done."
15. Impressions gained from an examination of industry journals and correspondence. See also House, Committee on Public Works, *National Highway Study*, p. 2.
16. R. A. Haber, Chief Engineer, Delaware, to Francis V. du Pont, April 6, 1953, BPR File 740.1.1 General June 1955–April 1953; Francis V. du Pont to Caleb Boggs, May 25, 1953, BPR File 740.1.1 General April 1954–January 1953, both Acc. No. 58-A-778, FRC; Evans, "The Great Highway Robbery—Or Is It?" pp. 65, 106, 110. See also Report for Files; Notes on the AASHO–AGC Joint Cooperative Meeting, October 1, 1953, Asheville, North Carolina, in Papers of the American Association of State Highway Officials–Associated General Contractors of America, Incorporated, Joint Cooperative Committee.
17. See statement of Governor Hugh Gregg of New Hampshire, June 10, 1953, in *National Highway Study*, II, pp. 242–243; State of Michigan, Senate Resolution No. 44, in *ibid.*, p. 244. S. 219 and H.R. 3637 provided for creation of a Highway Trust Fund.
18. Relationship of A.A.A. to National Highway Users Conference, c. February 13, 1953; see also Discussion of Relationship Between A.A.A. and Other Organizations in the Highway Field, February 13, 1953. For a review of the activities of AAA executives in the politics of road building and truck taxation, see remarks of Edward G. Rockwell, E. Ray Cory, Stuart B. Wright, and Matthew C. Sielski at the AAA Round-Table Discussion, February 12, 1953, Washington, D.C., all in Holland Papers, Truman Library.
19. Impressions gained from an examination of trucking industry journals; C. S. Morgan, "The Motor Transport Industry," in [Advisory Committee for the Transportation Study], National Resources Planning Board, *Transportation and National Policy*, p. 401; T. W. Van Metre, *Transportation in the United States* (Chicago: The Foundation Press, 1939), pp. 347, 350. See also Ellis W. Hawley, *The New Deal and the Problem of Monopoly: A Study in Economic Ambivalence* (Princeton: Princeton University Press, 1966), pp. 232–234.
20. David Beck, Roy A. Fruehauf, and Burge N. Seymour to Dwight D. Eisen-

hower, January 30, 1953, OF 122-N, Dwight D. Eisenhower Library, Abilene, Kansas.

21. Bureau of the Budget, Office of Legislative Reference, Legislative Program: Check List of Selected Proposals, November 14, 1953, OF 99-Z; Arthur F. Burns to the President, August 11, 1953, Papers of Arthur F. Burns; Walter Williams, Under Secretary of Commerce, "Ever Seen a Statistic," Remarks scheduled for delivery to the 41st National Safety Congress, October 21, 1953, Chicago, Stephen Hess Files; Text of Address by Sherman Adams, Assistant to the President, to the American Municipal Association, December 2, 1953, New Orleans, John S. Bragdon Files; Office of the Under Secretary (of Commerce) for Transportation, "The Potential Use of Toll Road Development in a Business Depression," December, 1953, John S. Bragdon Papers; and "Federal Highway Policy," November 19, 1953, Bryce N. Harlow Files, all in Eisenhower Library.
22. "Planning of Public Works," Chapter 10, January 10, 1954; "Role of Government in Economic Progress," Chapter 1 (third draft), January 17, 1954, both in OF 99 G-7; William J. Cox and Richard M. Zettel, "Report of the Highway Study Committee to the Commission on Intergovernmental Relations" (draft for committee use), January 28, 1954, Records of the President's Commission on Intergovernmental Relations, all in Eisenhower Library.
23. John V. Lawrence to J. Harry McGregor, November 9, 1953, reprinted in U.S. Congress, Senate, Roads Subcommittee of the Committee on Public Works, *Hearings on S. 2859, S. 2982, S. 3069, and S. 3184, Federal-Aid Highway Act of 1954*, 83rd Congress, 2d Session, 1954, pp. 322–323; Statement by John V. Lawrence to the House Public Works Committee, February 16, 1954, in U.S. Congress, House, Committee on Public Works, *Hearings on H.R. 7678, H.R. 7818, H.R. 7841, H.R. 7124, H.R. 7207, H.R. 14, H.R. 1407, H.R. 3528, and H.R. 3529, Federal-Aid Highway Act of 1954*, 83rd Congress, 2d Session, 1954, p. 146; Policy Position of the NHUC Board of Governors, printed in *ibid.*; Statement by Walter F. Carey to the National Conference on Highway Financing, Sponsored by the Chamber of Commerce of the United States, Washington, D.C., December 11, 1953, Records of the President's Commission on Intergovernmental Relations, Eisenhower Library.

    Automobile industry officials avoided the gas tax issue and emphasized the importance of federal spending on roads for national purposes, coordination of highway construction, and as a way of satisfying demand created by motorists. See James Cope, Vice President of Chrysler Corporation, to Clarence E. Manion, Chairman of the President's Commission on Intergovernmental Relations, November 16, 1954, Records of the President's Commission on Intergovernmental Relations, Eisenhower Library.
24. *Public Papers of the Presidents of the United States; Dwight D. Eisenhower; Containing the Public Messages, Speeches, and Statements of the Presidents, 1954* (Washington, D.C.: U.S. Government Printing Office, 1960), pp. 15, 18, 176, 245.
25. Minutes of Meeting of Study Committee on Federal Aid to Highways, February 15–17, 1954, Records of the President's Commission on Intergovernmental Relations; Russell E. Singer to Sherman Adams, January 29, 1954, General File 158-A-1, both in Eisenhower Library; Statement of Al-

fred E. Johnson, President of AASHO and Chief Engineer of Arkansas, to the Senate Subcommittee on Roads of the Senate Committee on Public Works, February 19, 1954, in Senate, *Hearings on S. 2859 and Other Bills, Federal Aid Highway Act of 1954*, pp. 6–9; Hal H. Hale to the Executive Committee of the American Association of State Highway Officials, March 17, 1954, BPR File 740.1.1 General April 1954 (Acc. No. 58-A-778), FRC. See also Frank D. Merrill, Commissioner, New Hampshire Department of Public Works and Highways, to Sherman Adams, January 30, 1954, OF 141-B; Remarks of William J. Cox in Minutes of Meeting of Study Committee on Federal Aid to Highways, January 18–19, 1954, Records of the President's Commission on Intergovernmental Relations, both in Eisenhower Library.

26. U.S. Congress, House, Committee on Public Works, *Federal-Aid Highway Act of 1954*, 83rd Congress, 2d Session, 1954, H. Rept. No. 1308 to accompany H.R. 8127, pp. 1–2, 17, 20–22; U.S. *Congressional Record*, 83rd Congress, 2d Session, 1954, 100, Pt. 2, pp. 2849, 2851, 2854–2856, 2860. See also Representative John J. Dempsey of New Mexico to the President, March 2, 1954, OF 141-B, Eisenhower Library.

27. Arlyn E. Barnard, Executive Secretary, Maine Automobile Club, to Senator Margaret Chase Smith, March 12, 1954; Donald R. Belcher to Edward Martin, March 3, 1954, both in Records of the Senate Public Works Committee, Folder S. 2982, NA; Charles Marshall, President, Nebraska Farm Bureau Federation, to the President, March 5, 1954, BPR File 740.1.1 General April 1954 (Acc. No. 58-A-778), FRC; Roger W. Jones to I. Jack Martin, March 11, 1954, I. Jack Martin Files, Eisenhower Library. See also Sinclair Weeks to Edward Martin, March 12, 1954, BPR File 740.1.1 General April 1954 (Acct. No. 58-A-778), FRC.

28. U.S. Congress, Senate, Public Works Committee, *Federal-Aid Highway Act of 1954*, 83rd Congress, 2d Session, 1954, S. Rept. 1093 to accompany S. 3184, pp. 2–3, 8–10, 13–15, 17.

    For the views of government officials on public works spending, see Joseph M. Dodge, Director, Bureau of the Budget, to I. Jack Martin, Martin Files; Francis V. du Pont to Robinson Newcomb, March 24, 1954; [Neil H. Jacoby], "Measures Currently Available to the Federal Administration, Assuming an Accelerating Economic Decline during 1954 Calling for Maximum Counter-Cyclical Activity by Government," March 24, 1954, both in Bragdon Files, all in Eisenhower Library.

29. *Congressional Record*, 100, Pt. 4, pp. 4682, 4760–4761, 4784, 4789–4790.

30. Charles L. Dearing to Robert B. Murray, Under Secretary of Commerce for Transportation, April 13, 1954, Record Group 40, Department of Commerce, Records of the Office of the Secretary, Public Roads File (Acc. No. 56-A-468), Department of Commerce, Washington, D.C. (cited hereafter as Records of the Office of the Secretary of Commerce).

31. *Congressional Record*, 100, Pt. 4, pp. 5123–5124, 5126, 5146–5147, 5149.

32. See remarks of Governor Allan Shivers of Texas and Governor Dan Thornton of Colorado in [Executive Office of the President], *A Report on the Washington Conference of Governors*, April 26–28, 1954, edited for national security (Washington, D.C.: Executive Office Building, 1954), pp. 21–23.

5

THE HIGHWAY AND THE CITY, 1945–1955

1. Harold M. Mayer, "Moving People and Goods in Tomorrow's Cities," *The Annals of the American Academy of Political and Social Science* 242 (November, 1945), 116; George F. Emery, "Urban Expressways," *American Planning and Civic Annual,* ed. Harlean James (Washington, D.C.: American Planning and Civic Association, 1947), p. 127; John G. Marr, "Impact of Freeway Location upon Cities," paper to the California Chamber of Commerce, Central Coast Council, Oakland, California, February 13, 1948, BPR File 790 California 1949-1943, FRC; C. McKim Norton, "Metropolitan Planning," *Traffic Quarterly* 3 (October, 1949), 367; Cincinnati, The City Planning Commission, *The Cincinnati Metropolitan Master Plan,* 1948, pp. 79, 84.
2. Harland Bartholomew, "Development and Planning of American Cities," an address before the student body of the Carnegie Institute of Technology, May 10, 1949 (Pittsburgh: Carnegie Press Occasional Papers, Number One, April, 1950), p. 19; and "The Location of Interstate Highways in Cities," *American Planning and Civic Annual,* ed. Harlean James (Washington, D.C.: American Planning and Civic Association, 1949), p. 75.
3. T[heodore] J. Kent, "City and Regional Planning Needs in Relation to Transportation," *Proceedings of the First California Institute on Street and Highway Problems* (Berkeley: University of California, c. 1949), pp. 57–58. See also Robert B. Mitchell, Columbia University Institute for Urban Land Use and Housing Studies, "Coordination of Highway and City Planning," Highway Research Board, *Proceedings of the Twenty-eighth Annual Meeting,* eds. Roy W. Crum, et al. (Washington, D.C.: Highway Research Board, 1949), pp. 17–18.
4. Lubove, *Twentieth-Century Pittsburgh: Government, Business, and Environmental Change,* pp. 106–107, 111; Park H. Martin, Director of the Allegheny Conference on Community Development, "Pittsburgh's Golden Triangle," *American Planning and Civic Annual, 1951,* ed. Harlean James (Washington, D.C.: American Planning and Civic Association, 1951), p. 139.
5. Lubove, *Twentieth-Century Pittsburgh: Government, Business, and Environmental Change,* pp. 108, 110–111, 127.
6. Rice, "Roads Were Built for Commerce, Not Sightseeing," pp. 28–29, 114, 116, 118; MacCleery, "The Place of the Motor Vehicle in Post-War Transportation," pp. 65, 67–68, 70–72; National Highway Users Conference, *Dedication of Special Highway Revenues to Highway Purposes: An Analysis of the Desirability of Protecting Highway Revenues through Amendments to State Constitutions,* p. 5; NHUC, *The Eastman Report Finds That Highway Users Pay Their Way and More,* p. 16; Summary of Minutes of Joint Meeting, Board of Governors and Administrative Committee, National Highway Users Conference, December 14, 1949, October 11, 1951; Butler, "Our Highway Problem and Some Recommendations for Meeting It"; Burton H. Behling, "Summation of Remarks at National Conference on

Highway Financing," December 11, 1953, Records of the President's Commission on Intergovernmental Relations, Eisenhower Library.

7. Frank C. Balfour, "Acquisition of Access Rights in California," Highway Research Board, *Proceedings of the Twenty-fourth Annual Meeting* (unassembled), eds. Roy W. Crum and Fred Burggraf (Washington, D.C.: Highway Research Board, c. 1945), pp. 16, 21–22.
8. B[ertram] D. Tallamy, "Meeting the Urban Thoroughfare Challenge," *American Association of State Highway Officials 33rd Annual Convention, 1947* (Washington, D.C.: AASHO, 1947), pp. 155, 159. See also Donald Baker, "Financing Express Highways in Metropolitan Areas," *The American City* 61 (October, 1946), 93; Charles M. Noble, "Highway Planning in Metropolitan Areas," *American Planning and Civic Annual,* ed. Harlean James (Washington, D.C.: American Planning and Civic Association, 1948), p. 112; Harold Eckhardt, "Traffic Bottlenecks in Cities," *Proceedings of the Ohio Highway Engineering Conference, 1949* (Columbus: The Ohio State University, 1949), p. 219; K. A. MacLachlan, "Engineering and Economic Justification for Major Urban Transportation Improvements and Value of Origin-Destination Surveys," *Proceedings of the First California Institute on Street and Highway Problems,* p. 205.
9. Summary of discussion of paper presented by K. A. MacLachlan, in *ibid.,* pp. 211–212.
10. Public Roads Administration, "Explanation and Interpretation of the Federal-Aid Highway Act of 1944"; PRA, "Federal Aid Highway Act of 1944."
11. Herbert S. Fairbank, "The Federal-Aid Highway Act—A Promise and a Challenge to Cities," paper before a joint meeting of the Engineers Society of Milwaukee and the Wisconsin Section of the American Society of Civil Engineers, April 18, 1945, BPR File 740.1.1 1946-45 General, FRC. See also David R. Levin, Public Roads Administration, "Legislative and Administrative Implementation of the Post-War Highway Program," Highway Research Board, *Proceedings of the Twenty-fourth Annual Meeting* (unassembled), pp. 7–8, 12–13.
12. Summary of remarks of Lawrence I. Hewes to the Commonwealth Club of California, December 7, 1945, BPR File 790 California 1949-43, FRC; and his "Metropolitan Freeways and Mass Transportation," *Transactions of the Commonwealth Club of California* 40 (San Francisco, 1946), 101, 103. See also Thomas H. MacDonald, "The Case for Urban Expressways: Long-Range Planning of Adequate Highway Facilities Will Save Many Cities from Stagnation and Decay," *The American City* 62 (June, 1947), 92; Joseph Barnett, Public Roads Administration, "Express Highway Planning in Metropolitan Areas," *Proceedings of the American Society of Civil Engineers* 72 (March, 1946), 301–302; and Barnett, "Urban and Inter-City Road Improvement," *Proceedings of the Ohio Highway Engineering Conference* (Columbus: The Ohio State University, 1947), pp. 27, 34.
13. Richard O. Davies, *Housing Reform During the Truman Administration* (Columbia, Missouri: University of Missouri Press, 1966), p. 40.
14. Material Proposed for Inclusion in the State of the Union Message of the President, November 14, 1947, Clark Clifford Papers, Truman Library.

15. Notes on Title 1—Slum Clearance—of Proposed Housing Bill, December 21, 1948, BPR File 740.1.1 General 1949-48, FRC.
16. Harry S. Truman to Philip B. Fleming, April 20, 1949, BPR File 740.1.1 1949-48, FRC.
17. See Scott, *American City Planning since 1890*, pp. 464–466, 502, 538–539; James Q. Wilson, ed., *Urban Renewal: The Record and the Controversy* (Cambridge: The MIT Press, 1966), pp. xv–xvi. See also Blake McKelvey, *The Emergence of Metropolitan America, 1915–1966* (New Brunswick: Rutgers University Press, 1968), pp. 134–135, 169.

    Scott argues that Housing and Home Finance Agency officials "inevitably would exercise a certain pressure to improve the techniques of planning." But the predicted intervention was imperceptible. Compare Scott, *American City Planning since 1890*, p. 466 with Wilson, *Urban Renewal*, pp. xv–xvi.
18. Jeanne R. Lowe, *Cities in a Race—With Time: Progress and Poverty in America's Renewing Cities* (New York: Random House, 1967), pp. 410–415.

    Two political scientists completed careful studies of renewal decision making in New Haven. See Robert A. Dahl, *Who Governs? Democracy and Power in an American City* (New Haven: Yale University Press, 1961), and Nelson W. Polsby, *Community Power and Political Theory* (New Haven: Yale University Press, 1963). Alan R. Talbot, *The Mayor's Game: Richard C. Lee of New Haven and the Politics of Change* (New York: Harper and Row, 1967), is a lucid but uncritical account of Mayor Lee's role in renewal; Fred Powledge, *Model City, a Test of American Liberalism: One Town's Efforts to Rebuild Itself* (New York: Simon and Schuster, 1970), surveys some of the results of renewal and finds them insufficient or counterproductive.
19. Polsby, *Community Power and Political Theory*, p. 71; Dahl, *Who Governs? Democracy and Power in an American City*, p. 117; Scott, *American City Planning since 1890*, p. 530; and remarks of Martin Meyerson to the 1956 meeting of the American Institute of Planners as cited in *ibid.*, p. 529. Talbot, *The Mayor's Game: Richard C. Lee of New Haven and the Politics of Change*, pp. 18–19.
20. *Ibid.*, p. 19.
21. See Lowe, *Cities in a Race—With Time*, pp. 422, 433–434, 467–468; Talbot, *The Mayor's Game: Richard C. Lee of New Haven and the Politics of Change*, pp. 105–106, 144–145; Wilfred Owen, *Cities in the Motor Age* (New York: The Viking Press, 1959), p. 36.
22. Alan Altshuler, *The City Planning Process: A Political Analysis* (Ithaca, New York: Cornell University Press, 1965), p. 75.
23. *Ibid.*, pp. 46, 48.
24. *Ibid.*, pp. 45–46.
25. *Ibid.*, pp. 40, 48, 73.
26. A. Theodore Brown, *The Politics of Reform: Kansas City's Municipal Government, 1925–1950* (Kansas City: Community Studies, Inc., 1958), p. 389; BPR, *Highway Statistics: Summary to 1955*, p. 74.

6

DWIGHT D. EISENHOWER AND EXPRESS HIGHWAY POLITICS, 1954–1955

1. "The Rate of Economic Progress," c. November, 1953, pp. 5–6; "The Administration Program for Economic Expansion," June 9, 1954, pp. 1–3, both in Records of the Office of the Council of Economic Advisers; Memorandum for Governor Adams, February 19, 1954, Gabriel Hauge Files; Robert B. Murray, Jr., to Gabriel Hauge, Assistant to the President, December 23, 1953, Burns Papers; Francis V. du Pont to Robinson Newcomb, March 24, 1954; John S. Bragdon to the Record, April 9 and April 12, 1954, all in Bragdon Files; Roger W. Jones to Gerald D. Morgan, April 12, 1954, OF 141-B, all in Eisenhower Library.
2. Bragdon for the Files, April 12, 1954; Dwight D. Eisenhower to Governor Adams, May 11, 1954, Administrative File; Informal Remarks of the President to the White House Conference on Highway Safety, February 12, 1954, Speech File, both in Eisenhower Library.
3. "The Administration Program for Economic Expansion," pp. 1–3. See also Robinson Newcomb, "How We Might Get the Safe and Adequate Highways We Need," p. 5, Papers of Neil J. Jacoby, Eisenhower Library.
4. [John S. Bragdon], Comparison of Moses–Tallamy Plan and Council of Economic Advisers' Draft, May 18, 1954; John S. Bragdon to Arthur F. Burns, May 24, 1954, both in Bragdon Files, Eisenhower Library.
5. Bertram D. Tallamy and Robert Moses to Sherman Adams, May 4, 1954, Bragdon Files, Eisenhower Library.
6. Bertram D. Tallamy and Robert Moses, Draft of a Bill to Create a Continental Highway Finance Corporation, c. May 1, 1954, Bragdon Files, Eisenhower Library; Tallamy and Moses to Adams.
7. Bragdon to Burns, May 24, 1954; Tallamy and Moses, Draft of a Bill to Create a Continental Highway Finance Corporation. See, for examples of on-going disputes, John S. Bragdon for the Record (Conference with Meyer Kestnbaum), June 11, 1954, Bragdon Files; Robert B. Murray, Jr., to Richard M. Zettel, July 2, 1954, Records of the President's Commission on Intergovernmental Relations, both in Eisenhower Library.
8. Sherman Adams to the Director, Bureau of the Budget, May 11, 1954; John S. Bragdon to Arthur F. Burns, June 11 and June 25, 1954; Arthur F. Burns to Sherman Adams, June 17, 1954, all in Bragdon Files, Eisenhower Library.
9. Remarks of Vice President Richard M. Nixon to the Governors' Conference, June 12, 1954, OF 147 A-1, Eisenhower Library. Nixon delivered his remarks from Eisenhower's text. Eisenhower did not attend the conference due to a death in his wife's family.
10. Arthur F. Burns to the President, July 22, 1954, Burns Papers. Members of Congress were not informed of the proposed committee's scope, organization, and membership. Compare Senator Francis Case to Sherman Adams, July 17, 1954; Representative John J. Dempsey to the President, July 23, 1954; J. Harry McGregor to Francis V. du Pont, July 31, 1954; Sherman Adams to John J. Dempsey, August 4, 1954, all OF 141 B-1, with John S.

Bragdon for the Record, July 21, 1954, Bragdon Files, and Burns to the President, July 22, 1954, all in Eisenhower Library.

11. Dwight D. Eisenhower to George M. Humphrey, August 20, 1954, OF 141 B-1-A; Dwight D. Eisenhower to Sinclair Weeks, August 20, 1954; Minutes of Meeting of Interagency Committee on the President's Highway Program, September 9, 1954, both in Bragdon Files; Lucius D. Clay, interview on February 20, 1971, conducted by Edward Edwin, Columbia University, for the Dwight D. Eisenhower Oral History Project, p. 106, all in Eisenhower Library. See also Arthur Minnich to Thomas E. Stephens, August 23, 1954, OF 141 B-1-B; Francis V. du Pont to Roemer McPhee, September 9, 1954, OF 141 B-1-A, both in Eisenhower Library.
12. Clay and Adams selected Stephen D. Bechtel, a construction contractor; William A. Roberts, President, Allis Chalmers Manufacturing Company; David Beck, head of the Teamsters Union; and Sloan Colt, President, Bankers Trust Company of New York. See Clay interview, February 20, 1971, p. 102; Lucius D. Clay to the President, August 30, 1954, OF 141 B-1-B, Eisenhower Library; inferences drawn from Clay's description of their choices in Clay to the President. For names of others recommended for membership on the committee, see correspondence in Organization File of the President's Advisory Committee on a National Highway Program, Eisenhower Library.
13. Robinson Newcomb to the Council [of Economic Advisers], August 23, 1954, Bragdon Files, Eisenhower Library.
14. Bertram D. Tallamy, et al., "A Proposal for the Financing and Administration of the National System of Interstate Highways and Other Highway Responsibilities of the Federal Government," c. July 31, 1954, Records of the President's Advisory Committee on a National Highway Program, Eisenhower Library.
15. John S. Bragdon to Arthur F. Burns, August 3 and September 8, 1954; Minutes of Meeting of Interagency Committee on the President's Highway Program, September 9, 1954, all in Bragdon Files; John S. Bragdon to the Council, September 9, 1954, Burns Papers, all in Eisenhower Library.
16. Minutes of Meeting of Interagency Committee on the President's Highway Program, September 9, 1954.
17. John S. Bragdon to Arthur F. Burns, October 26, 1954, Burns Papers; Harold L. Pearson to the Files, October 28, 1954, Bragdon Papers; John S. Bragdon to the Record, October 27, 1954; [Budget Bureau], Suggested Principles for National Highway System, October 27, 1954; John S. Bragdon, Note for National Highway System, n.d., all in Bragdon Files, all in Eisenhower Library. For an indication of the reluctance of budget, treasury, and CEA officials to commit themselves to aspects of highway financing, see Bragdon to the Record, October 27 and 28, 1954, Bragdon Files, Eisenhower Library.
18. Minutes of Meeting of Interagency Committee on the President's Highway Program, September 9 and 15, 1954; John S. Bragdon to the Council [of Economic Advisers], September 27, 1954, both in Bragdon Files; John S. Bragdon to the Council [of Economic Advisers], November 10, 12, and 23, 1954; Arthur F. Burns, "Preliminary Thoughts on Economic Legislation," December 3, 1954 (unused draft), last four items in Burns Papers; John S.

Bragdon to the Council [of Economic Advisers], November 2, 1954; John S. Bragdon to the Council [of Economic Advisers], "The Role of Public Works in Maintaining Economic Stability and Growth," December 28, 1954, both in Jacoby Papers; George M. Humphrey, Remarks of George M. Humphrey to the National Governors' Conference, April, 1954, pp. 58–59. See also Raymond J. Saulnier to the Council [of Economic Advisers], December 9, 1954, Burns Papers.

19. Memorandum for File, November 19, 1954; Minutes of Meeting of Interagency Committee on the President's Highway Program, November 19, 1954, both in Bragdon Files, Eisenhower Library.

20. Gordon Keith to the Council [of Economic Advisers], October 18, 1954, Bragdon Files; [National Grange], "Report of the Committee on Transportation," c. December, 1954; Matt Triggs, American Farm Bureau Federation, to Lucius D. Clay, December 20, 1954; Russell E. Singer, AAA, to Lucius D. Clay, October 28, 1954, last three items in Records of the President's Advisory Committee on a National Highway Program, all in Eisenhower Library.

    See also Chamber of Commerce of the United States, Construction and Civic Development Department, Subcommittee on Highway Development, "Suggested Recommendations for National Chamber Policy with Respect to the President's Highway Proposal," November 30, 1954, Burns Papers; Governors' Conference, Special Committee on Highways, "Suggested Program as Recommended by the Governors' Conference Special Committee on Highways," November 9, 1954, Records of the President's Advisory Committee on a National Highway Program, both in Eisenhower Library.

21. Memo in files entitled Clay Comments, October 27, 1954; Notes on General Clay's Discussion of December 18, 1954, both in Records of the President's Advisory Committee on a National Highway Program; John S. Bragdon for the Record, December 13, 1954, Bragdon Files, all in Eisenhower Library; President's Advisory Committee on a National Highway Program, *A Ten-Year National Highway Program: A Report to the President* (n.p., January, 1955), pp. 21–22, 28.

22. Papers of James C. Hagerty, Hagerty Diary, December 3 and 13, 1954; Dwight D. Eisenhower to Lucius D. Clay, January 26, 1955, Administrative File, all in Eisenhower Library.

23. Gabriel Hauge to Sherman Adams, December 23, 1954, OF 141 B-1-B; Council of Economic Advisers to the White House, January 21, 1955; John S. Bragdon to the Council [of Economic Advisers], January 20, 24, 26, and 27, 1955, all in Bragdon Files; Harold L. Pearson, Draft of Financing Proposals, January 25, 1955, Records of the President's Advisory Committee on a National Highway Program, all in Eisenhower Library.

    Bragdon subordinated many of his own views on toll financing, management, and mileage in order to participate in the CEA letter to the White House, January 21, 1955. Compare *ibid.* and Bragdon's draft of a letter to the White House, January 20, 1955, Burns Papers. In formal meetings, however, Bragdon urged support for Clay's program. See Bragdon to the Council, January 20, 1955.

    Between December, 1954, and February, 1955, members of the Highway Study Committee of the President's Commission on Intergovernmental Relations continued to squabble about the scope and direction of an updated federal road program. See correspondence in Records of the President's Commission on Intergovernmental Relations; see also Frank D. Merrill to Sherman Adams, February 17, 1955, GF 158 A-1, all in Eisenhower Library.

24. Bragdon for the Record, December 13, 1954; John S. Bragdon to the Council [of Economic Advisers], February 1, 1955, Bragdon Files; Francis C. Turner to Kevin McCann, January 28, 1955; Lucius D. Clay to Governor Paul Patterson, February 17, 1955, both in Records of the President's Advisory Committee on a National Highway Program, all in Eisenhower Library; *Public Papers of the Presidents of the United States: Dwight D. Eisenhower, 1955* (Washington, D.C.: U.S. Government Printing Office, 1959), p. 280.

    In drafting the administration bill, treasury and budget officials assumed responsibility for details of financing while du Pont handled remaining sections. But even this division of responsibilities did not eliminate disputes between interagency members. See Bragdon to the Council, January 24, 1955; and John S. Bragdon to Arthur F. Burns, February 7, 10, and 17, 1955; John S. Bragdon to Francis V. du Pont, February 9, 1955, all in Bragdon Files, Eisenhower Library.

25. John V. Lawrence to the President; Arthur M. Hill to the President; Andrew J. Sardoni to the President, all dated January 5, 1955; Andrew J. Sardoni to Dwight D. Eisenhower, January 17, 1955, all in GF 158 A-1; John S. Bragdon for the Record, May 4, 1955; Report on Proposed Highway Policies Prepared by the Construction and Civic Development Department of the United States Chamber of Commerce, c. March, 1955, both in Bragdon Files; Francis V. du Pont to I. Jack Martin, February 25, 1955, Martin Files, all in Eisenhower Library; Statement of K. B. Rykken to the National Conference on Highway Financing Sponsored by the Chamber of Commerce of the United States, Washington, D.C., January 13–14, 1955; The Clay Committee Report in Relation to AAA Highway Policies, c. January, 1955, both in Holland Papers, Truman Library; Governor Abraham Ribicoff to Senator Prescott Bush, March 15, 1955, BPR File 740.1.1 H.J. Res. 113 (Acc. No. 58-A-778), FRC; Robert T. Stevens to Dennis Chavez, March 23, 1955, Papers of the Senate Public Works Committee, File 84A-E14, NA. See also Tucker, Anthony and Company, "Highlights from the Washington Conference to Discuss the Clay Committee $101 Billion Road Project to Be Made to the Congress by President Eisenhower on January 27, 1955," p. 6 (Sponsored by the Chamber of Commerce of the United States); Frank D. Merrill to Sherman Adams, April 22, 1955, both in GF 158 A-1, Eisenhower Library; Light B. Yost, "An Investment in Progress," *University of Tennessee Record* 58 (July, 1955), 10–11.

    For evidence that administration officials were aware of views of industry leaders, see [John S. Bragdon], Memorandum to the Council [of Economic Advisers], January 20, 1955, Jacoby Papers, Eisenhower Library.

26. Memorandum of Conversation with General Clay, February 7, 1955, Administrative File; Hagerty Diary, February 16, 1955, both in Eisenhower Library.
27. Hagerty Diary, February 21, 1955, Eisenhower Library.
28. Bragdon for the Record, May 4, 1955; John S. Bragdon for the Record,

April 29, 1955, Bragdon Files; J. Harry McGregor to George M. Humphrey, April 18, 1955, OF 141 B-1; Governor Paul Patterson to Lucius D. Clay, February 15, 1955, Records of the President's Advisory Committee on a National Highway Program, all in Eisenhower Library.

29. Harry F. Byrd, "Statement by Senator Harry F. Byrd Relative to Clay Commission Highway Report," *Virginia Municipal Review* (January, 1955), 10.
30. Hagerty Diary, February 16 and 21, 1955; Francis V. du Pont to Harmer E. Davis, February 28, 1955, BPR File 481 California FAS General (Acc. No. 58-A-778), FRC; McGregor to Humphrey; J. Harry McGregor to Sherman Adams, April 1, 1955, OF 141 B-1, Eisenhower Library.
31. John S. Bragdon to the Council [of Economic Advisers], March 31, 1955, Bragdon Files, Eisenhower Library.
32. *Ibid.*; John S. Bragdon to the Council [of Economic Advisers], April 4, 1955; John S. Bragdon to Arthur F. Burns, April 11, 1955; W. D. Gradison to J. E. Reeve, April 7, 1955; J. E. Reeve to H. L. Pearson, April 7, 1955, all in Bragdon Files, Eisenhower Library. See also Office of the Secretary of the Treasury, Analysis Staff, Debt Division, "Financing the Highway Program through Tolls," April 15, 1955; and see as well Bragdon's critique in John S. Bragdon to the Council [of Economic Advisers], April 22, 1955, both in Bragdon Files, Eisenhower Library.
33. Bragdon to Burns, April 11, 1955; John S. Bragdon to Arthur F. Burns, April 12 and 22, 1955; John S. Bragdon to the Council [of Economic Advisers], May 4, 1955, all in Bragdon Files, Eisenhower Library.
34. Comments on S. 1160, n.d., Records of the Office of the Chief of Transportation, File RCSI #7 (d), FRC.
35. Remarks of General Lucius D. Clay to the Washington Conference of Governors, May 2, 1955, OF 147 A-2, Eisenhower Library. See also Bragdon for the Record, April 29, 1955; Remarks of Francis V. du Pont to the Washington Conference of Governors, May 2, 1955, OF A-2, Eisenhower Library; du Pont to Davis.

    Early in 1955, du Pont resigned as commissioner of the Bureau of Public Roads to serve as congressional liaison for highway matters under Secretary of Commerce Sinclair Weeks.
36. Bragdon noted the 8-4 tally in a memo to Arthur F. Burns, May 10, 1955, Bragdon Files; U.S. Congress, Senate, Committee on Public Works, *Federal-Aid Highway Act of 1955*, 84th Congress, 1st Session, S. Rept. No. 350 to accompany S. 1048, pp. 8–10, 13–14, 19–20, 22–29; George D. Riley to Dennis Chavez, February 21, 1955, Records of the Senate Committee on Public Works, File 84A-E14, NA; "Comparison of Administration and Gore Highway Bills," n.d., OF 141-B, Eisenhower Library.
37. *Congressional Record*, 84th Congress, 1st Session, 1955, 101, Pt. 5, pp. 6976, 7033.
38. H.R. 7072 published in U.S. Congress, House, Committee on Public Works, *Hearings, National Highway Program*, Part 2, 84th Congress, 1st Session, 1955, pp. 1097–1100.
39. Edward Margolin to Louis S. Rothschild, October 25, 1955, Records of the Office of the Secretary of Commerce, File LSR-PR Interstate System—general, Department of Commerce; Robinson Newcomb to Arthur F. Burns, August 8, 1955, Martin Files; Neil J. Curry, "What's Ahead in Truck

Transportation?" Address to the 32nd Annual Meeting of the Associated Traffic Clubs of America, September 20, 1955, Howard Pyle Files; Dave Beck to Lucius D. Clay, February 20, 1956, GF 158 A-1, last three in Eisenhower Library.

40. Remarks of Secretary of the Treasury George M. Humphrey to the House Committee on Public Works, in *Hearings, National Highway Program,* Part 2, pp. 1188–1189; Dwight D. Eisenhower Press Conference, June 29, 1955, Press Conference File. See also Philip A. Ray to I. Jack Martin, July 1, 1955, Martin Files; Collis Stocking to the Council [of Economic Advisers], prepared by John S. Bragdon, May 27, 1955, Records of the Office of the Council of Economic Advisers, all in Eisenhower Library.
41. General Counsel to the Under Secretary for Transportation, July 6, 1955, Records of the Office of the Secretary of Commerce, Interstate System—Legislative Comments File (Acc. No. 56-A-468), Department of Commerce; U.S. Congress, House, Committee on Public Works, *Federal-Aid Highway Act of 1955,* 84th Congress, 1st Session, H. Rept. No. 1336 to accompany H.R. 7474, pp. 1–2, 9; Newcomb to Burns.
42. *Congressional Record,* 101, Pt. 9, pp. 11561, 11695, 11709–11710. Those close to the trucking and congressional scenes observed much the same pattern. See, for example, Newcomb to Burns; Curry, "What's Ahead in Truck Transportation?"; U.S. Department of Commerce, Bureau of Public Roads, *Development of the Interstate Highway System* (Washington, D.C.: U.S. Department of Commerce, Bureau of Public Roads, 1964), p. 4; Henry J. Kaltenbach, "Proposed Federal Legislation for Highways," *Proceedings of the Forty-first Annual Meeting of the American Association of State Highway Officials* (Washington, D.C.: American Association of State Highway Officials, c. 1956), p. 53.
43. *Congressional Record,* 101, Pt. 9, pp. 11717–11718; Louis Shere to the Council [of Economic Advisers], August 1, 1955, Bragdon Papers, Eisenhower Library.
44. Dwight D. Eisenhower, *Mandate for Change, 1953–1956: The White House Years* (Garden City: Doubleday and Company, 1963), pp. 501–502; *Public Papers of the Presidents of the United States: Dwight D. Eisenhower, 1955,* p. 763.
45. "Legislative Program for 1955," May 5, 1955, p. 23, Records of the Office of the Council of Economic Advisers, Eisenhower Library; Hagerty Diary, February 21, 1955.

7

THE INTERSTATE HIGHWAY ACT OF 1956

1. John S. Bragdon to Colonel Andrew J. Goodpaster, July 28, 1955, OF 141-B, Eisenhower Library.
2. John S. Bragdon to Sherman Adams, September 27, 1955, Martin Files, Eisenhower Library.
3. Cabinet Paper, October 1, 1955 (for September 30, 1955), Records of the

White House Cabinet Secretariat, Eisenhower Library. Members of the Cabinet Committee on the highway program designated themselves the Presidential Advisory Committee.

4. Jack Martin to Sherman Adams, October 18, 1955; Jack Martin to Howard Pyle, October 19, 1955, both in OF 141-B, Margolin to Rothschild. See also Frank D. Merrill to Sherman Adams, October 24, 1955, OF 141 B-1, all in Eisenhower Library.
5. John S. Bragdon to Arthur F. Burns, November 2, 1955; John S. Bragdon to Sherman Adams, November 3, 1955, both in Bragdon Files; John S. Bragdon to Marion B. Folsom, October 5, 1955; Minutes of First Meeting of Advisory Committee on Federal Public Works, October 18, 1955, both in Records of the White House Cabinet Secretariat; Gabriel Hauge to the Secretary of Commerce, November 1, 1955, OF 141-B. See also Raymond J. Saulnier to Gabriel Hauge, August 30, 1957, OF 114, all in Eisenhower Library.
6. Cabinet Paper, October 1, 1955 (for September 30, 1955); [John S. Bragdon], Notes on Possible Federal Highway Legislation, November 30, 1955, Bragdon Files, Eisenhower Library.
7. J. Harry McGregor to Jack Martin, November 4, 1955, Martin Files; [Bragdon], Notes on Possible Federal Highway Legislation; [John S. Bragdon], Memorandum for the Record, December 2, 1955, Bragdon Files; [Council of Economic Advisers], "Enlarging Our Public Assets and Developing Our Resources," December 13, 1955, Records of the Office of the Council of Economic Advisers, all in Eisenhower Library; Fred Schwengel to Karl LeCompte, November 9, 1955, Frederick E. Biermann Collection, University of Iowa, Iowa City, Iowa.
8. Henry W. Osborne to John S. Bragdon and Colonel Meek, November 17, 1955, Bragdon Files, Eisenhower Library.

   See also Karl LeCompte to Fred Schwengel, November 18, 1955, Biermann Collection, University of Iowa; Frazer B. Wilde to Sherman Adams, November 23, 1955, OF 141-B; J. Harry McGregor, "Adequate Highways —How and When," c. January, 1956, p. 2, GF 158 A-1, both in Eisenhower Library.
9. *Congressional Record*, 84th Congress, 2d Session, 1956, 102, Pt. 5, pp. 7110, 7116–7118, 7146; *ibid.*, Pt. 6, pp. 7221–7222.

   The Boggs and Fallon bills were consolidated into one, H.R. 10660.
10. *Ibid.*, Pt. 5, pp. 7119–7121, 7128, 7132, 7134.
11. Eisenhower Diary, Telephone Calls, January 16 and 31, 1956; L. A. Minnich, Jr., to Rowland L. Hughes, January 31, 1956, Eisenhower Diary, all in Eisenhower Library.
12. Remarks of Jess N. Rosenberg to the Commonwealth Club of California, February 20, 1956, *Transactions of the Commonwealth Club of California* 100 (February, 1956), p. 78; William Noorlag, Jr., to Representative Sidney R. Yates, February 2, 1956; James D. Mann to Representative Jere Cooper, March 13, 1956; Central Motor Freight Association, *Newsgram*, April 2, 1956; P. M. Greenberg to Sidney R. Yates, April 6, 1956, all in Papers of Sidney R. Yates, Truman Library.
13. *Congressional Record*, 102, Pt. 5, pp. 7117, 7121, 7148; *ibid.*, Pt. 6, pp.

7178–7179, 7181–7182. Reed was Chairman of the House Ways and Means Committee.

14. *Ibid.*, Pt. 5, pp. 7117–7118; *ibid.*, Pt. 6, 7180–7181, 7184. See also Margolin to Rothschild.

    For views of members of Congress on paying compensation to state toll authorities, see U.S. Congress, House, Committee on Public Works, *Hearings on H.R. 8836, National Highway Program, Federal-Aid Highway Act of 1956*, 84th Congress, 2d Session, 1956, pp. 13–16, 27–28, 33–34, 37. In 1956, Commerce Department officials backed away from reimbursement, citing additional costs. See remarks of Secretary of Commerce Weeks in *ibid.*, p. 11.

    Members of the administration also thought it would prove advantageous to postpone or exclude certain items. See Memorandum for Governor Adams, April 10, 1956, Martin Files, Eisenhower Library.

15. *Congressional Record*, 102, Pt. 6, pp. 7180–7184; Kaltenbach, "Proposed Federal Legislation for Highways," p. 53.

16. Associated General Contractors of America, Reports of Meetings of AASHO–AGC Joint Cooperative Committee, January 19, 1956, p. 2, in AASHO–AGC Papers, Washington, D.C.; Allen Early, Jr., to Representatives Walter Rogers and Martin Dies and Senators Lyndon B. Johnson and Price Daniel, March 1, 1956, Records of the Senate Public Works Committee, File 84A-E14, NA; Memorandum for Governor Adams, April 10, 1956; L. A. Minnich, Jr., to Percival F. Brundage, April 24, 1956, Eisenhower Diary, Eisenhower Library. See also correspondence in Records of the Senate Public Works Committee, File H.R. 10660, NA.

    For congressional views of Davis-Bacon, see *Congressional Record*, 102, Pt. 6, pp. 7185–7206. For the views of a trade association leader in the road construction industry, see House, Committee on Public Works, *Hearings on H.R. 8836, National Highway Program, Federal-Aid Highway Act of 1956*, pp. 266–293.

17. Lyndon B. Johnson to Harry F. Byrd, March 19, 1956; George M. Humphrey to Harry F. Byrd, March 23, 1956, both in Record Group 50, Department of the Treasury, Records of the Secretary of the Treasury, Highway Program File, NA.

    See also Legislative Leadership Meeting, Supplementary Notes, January 31, 1956; L. A. Minnich, Jr., to Percival F. Brundage, May 22, 1956, both in Eisenhower Library; Minnich, Jr., to Brundage, April 24, 1956; Percy Rappaport to Jack Martin, c. April 7, 1956, Martin Files, all in Eisenhower Library.

    For language of the Fallon-Boggs bill obligating the treasury for highway construction expenses, see *Congressional Record*, 102, Pt. 6, pp. 7183–7184. For estimates of Trust Fund deficits, see *ibid.*, Pt. 5, p. 7153.

    Between February and late May, 1956, members of the House, state road engineers and governors, and Department of the Army transport officers sent along special requests to members of the Senate. Often, senators petitioned one another. Some sought revisions of the apportionment formula; others, especially engineers, wanted to eliminate hearings on by-pass decisions and block compensation payments to tenants forced to move; still others were promoting connection of toll and Interstate roads, federal guarantee of bond issues, or closer control by Congress of the Bureau of

Public Roads. See correspondence in Records of the Senate Public Works Committee, Files 84A-E14 and H.R. 10660, both in NA; see also Memo of meeting of D. K. Chacey, General Browning, et al., May 28, 1956, in Records of the Office of the Chief of Transportation, File RCSI #7 (d), FRC; Memo from Governor Robert F. Kennon, March 12, 1956, GF 158 A-1, Eisenhower Library.

18. *Congressional Record,* 102, Pt. 7, pp. 9070, 9075, 9080, 9117, 9232–9235, 9248. Members of the Senate Finance Committee voted to increase the surcharge to $2.50 per thousand pounds over 26,000, allowing a small savings to truckers with units close to the limit.

19. John C. Stennis to Dennis Chavez, June 1, 1956; telegrams in Senate Public Works Committee Files; Robert F. Wagner to Dennis Chavez, June 6, 1956, all in Records of the Senate Public Works Committee, H.R. 10660 File, NA; L. A. Minnich, Jr., to Percival F. Brundage, June 5, 1956, Eisenhower Diary, Eisenhower Library.

    See also Stuart Rothman to the Secretary [of Labor], June 1, 1956, Papers of James P. Mitchell, Eisenhower Library, for account of efforts of a Department of Labor official to influence conferees on Davis-Bacon.

20. "Comparison of H.R. 10660, as Passed by the House, with H.R. 10660, as Passed by the Senate," c. June, 1956, Records of the Senate Committee on Public Works, File H.R. 10660, NA; *Congressional Record,* 102, Pt. 8, pp. 10964, 10991–10997, 11004; Eisenhower, *Mandate for Change: The White House Years, 1953–1956,* p. 458.

21. J. L. Shotwell to D. W. Loutzenheiser, December 28, 1955, BPR Files: Primary Subjects and Symbols Yellow Reference Copies Administrator, subfile Engineering–Urban Highway Branch Loutzenheiser 1955-1958 (Acc. No. 62-A-1283), FRC; J. K. Crowson, "The Need for a Decision on Future Federal-Aid Policy," *Proceedings of the Forty-first Annual Meeting of the American Association of State Highway Officials,* p. 28.

22. For illustration of themes, see Curry, "What's Ahead in Truck Transportation?" p. 9; Charles M. Hayes, AAA, to Sidney R. Yates, February 13, 1956, Yates Papers, Truman Library; Sinclair Weeks to the Director, Bureau of the Budget, June 28, 1956, Records Officer Reports to President on Pending Legislation, Eisenhower Library; Central Motor Freight Association, *Newsgram.*

    See also Eisenhower Diary, January 31, 1956; John V. Lawrence, et al., "Statement of Highway Transportation Organizations on Urgent Need for Expanded Federal State Highway Program," c. 1956, Records of the Senate Public Works Committee, File 84A-E14, NA; Remarks of Senator Francis Case of South Dakota, June 26, 1956, in *Congressional Record,* 102, Pt. 8, p. 10964.

23. American Automobile Association, "Highlights of AAA Motorists Program for Better Highways," January 16, 1956, GF 158 A-1, Eisenhower Library; Eisenhower, *Mandate for Change: The White House Years, 1953–1956,* p. 548.

8

HIGHWAYS AND THE VALUES OF AMERICANS

1. George H. Deming to John S. Bragdon, October 3, 1956; John S. Bragdon to George H. Deming, October 5, 1956; John S. Bragdon to Arthur F. Burns, October 23, 1956, all in Bragdon Files, Eisenhower Library; Saulnier to Hauge.
2. John S. Bragdon to Robert Gray, December 23, 1958; Robert Gray to Robert E. Merriam, February 11, 1959; Robert Gray to Wilton B. Persons, March 31, 1960, all in Records of the White House Cabinet Secretariat; John S. Bragdon. "The Interstate Limited Access 90/10 Federal Aid System with Special Reference to Toll Financing and Intra-City Routing: Lost $30. $8. or $3.—How Many Billions," draft of manuscript dated June 15, 1961, pp. 3, 31, 33–34, Bragdon Papers, all in Eisenhower Library; Schwartz, "Urban Freeways and the Interstate System," pp. 446–447.
3. Scott, *American City Planning since 1890*, pp. 539–541; Thomas A. Morehouse, "Artful Interpretation: The 1962 Highway Act," in Michael N. Danielson, ed., *Metropolitan Politics: A Reader* (Boston: Little, Brown and Company, 1971), pp. 353–355; Owen, *Cities in the Motor Age*, pp. 32, 36, 38–39.

   Engineers also defined tenant relocation outside of their domain. See M. Justin Herman to Albert M. Cole, head of the Housing and Home Finance Agency, May 8, 1957, Record Group 207, Housing and Home Finance Agency, Records of the Office of the Administrator, NA.

   Curiously, E. H. Holmes, one of the engineers who had worked with MacDonald formulating Interstate plans, objected to the direction of the bureau's urban highway programming. See E. H. Holmes to C. D. Curtiss, May 3, 1956, BPR File Commissioner C. D. Curtiss 1955-1957, FRC.
4. For a fuller treatment of the popularity of the gasoline tax, see Burnham, "The Gasoline Tax and the Automobile Revolution."

   See also James J. Flink, "Three Stages of American Automobile Consciousness," *American Quarterly* 24 (October, 1972), 451–473, and his more critical *The Car Culture* for insightful analyses of the significance of the automobile for Americans.

Selected Bibliography

Neither government records in the National Archives nor private collections in presidential libraries are useful for locating accounts of day-to-day problems in urban areas and in the truck transport and highway construction industries. President Roosevelt's highway files, in fact, are sparse. Official word of traffic tangles often reached Presidents Truman and Eisenhower through economists and budget officers, but they sent data aggregated by economic sector and region, thus not in a form useful for pinpointing bottlenecks. Eisenhower's papers, though, do contain items affording a peek at presidential decision-making. Moreover, Lou E. Holland's papers at the Truman Library include about four boxes of American Automobile Association and Highway User Conference publications, many not available elsewhere.

Those who wish to study the process of organization building and the tensions of professionals operating in turbulent political and bureaucratic fields must turn to records in the National Archives of the Bureau of Public Roads, approximately five thousand boxes of fairly well organized material. BPR files also include maps, results of origin-destination surveys, and such, all useful for comparative analysis of traffic and road problems and the political response. *Toll Roads and Free Roads*, published by the bureau in 1939, is the most cogent statement of engineering logic applied to highway construction. Reports and papers of the National Resources Planning Board in the National Archives are a point of departure for the logic and goals of depression-era social scientists. What NRPB officials had in mind was application of the norms and findings of the Chicago School of Sociology to the national scene.

Conference proceedings—published annually by truckers and planners—provided a useful background for my own study. Speakers focused on daily routines and problems, and perceived solutions in terms celebrated by men inside their own profession and industry. Reports of commissions on city planning, churned-out by larger agencies, delineate the politics of members of a quasi profession. Roughly speaking, planners operated without a discrete body of information and agreed-upon norms of application, without access to political institutions. Harland Bartholomew, Theodore J. Kent, and Wilfred Owen were, in their writings, sensitive observers of planning, urban affairs, and highway construction. The proceedings of the Highway Research Board, of the American Association of State Highway Officials, and of highway engineering institutes contain critical comment as well as details of road finance and design. Moreover, officials of the Bureau of Public Roads, men with gigantic ambitions for American road building, wrote technical and popular articles.

Scholars of highway policy making have produced mostly tendentious accounts. What much of the secondary literature amounts to is an effort to blame state antidiversion legislation and the Highway Trust Fund for urban ills and the collapse of mass transit. Nonetheless, John C. Burnham's article on the origins of the gasoline tax remains the most perceptive and intelligent discussion of the social bases of the antidiversion impulse. Charles L. Dearing, *American Highway Policy*, contains a thoughtful defense of beneficiary payment. James J. Flink and John B. Rae, the senior scholars of American automotive history, point to the immense popularity of automobiles in the job, commuting, and recreational habits of Americans.

Overall, however, historians have not paid sufficient attention to highway matters. For the 1920s and 1930s, we require analyses of road politics at the state, county, and municipal levels. Studies of the upward shift in the locus of funding and authority—from township to state government—would prove valuable. While geographers and transport economists claim to know a great deal about highway impact, say, of road construction as a factor in plant location choices, historians have ignored highway building as a variable in changing social-spatial arrangements. If, indeed, politicians, economists, and the rest debate creation of countercyclical machinery in highway legislation, historians still have not made sense of the impact of huge federal transfer payments on aggregate or sector productivity and on federal-local relationships. Histories of truck operators and highway construction contractors—divided as they were by region, by scale of operations, and by levels of political self-consciousness—would make useful additions to our understanding of both American business organization and politics. In short, nearly every phase of American road building requires careful scrutiny by historical scholars.

PERSONAL AND BUSINESS MANUSCRIPT COLLECTIONS

Associated General Contractors of America, Incorporated–American Association of State Highway Officials. Papers of the Associated General Contractors of America, Incorporated–American Association of State Highway Officials' Joint Cooperative Committee. Associated General Contractors of America, Incorporated. Washington, D.C.

Biermann, Frederick E. Papers. The University of Iowa Library. Iowa City, Iowa.

Blough, Roy. Papers. The Harry S. Truman Library. Independence, Missouri.

Bragdon, John S. Papers. The Dwight D. Eisenhower Library. Abilene, Kansas.

Burns, Arthur F. Papers. The Dwight D. Eisenhower Library. Abilene, Kansas.

Clark, John D. Papers. The Harry S. Truman Library. Independence, Missouri.

Clifford, Clark M. Papers. The Harry S. Truman Library. Independence, Missouri.

Eisenhower, Dwight D. Papers. The Dwight D. Eisenhower Library. Abilene, Kansas.

Foley, Raymond M. Papers. The Harry S. Truman Library. Independence, Missouri.

Hagerty, James C. Papers. The Dwight D. Eisenhower Library. Abilene, Kansas.

Holland, Lou E. Papers. The Harry S. Truman Library. Independence, Missouri.

Hopkins, Harry L. Papers. The Franklin D. Roosevelt Library. Hyde Park, New York.

Jacoby, Neil J. Papers. The Dwight D. Eisenhower Library. Abilene, Kansas.

Mitchell, James P. Papers. The Dwight D. Eisenhower Library. Abilene, Kansas.
Nash, Philleo. Papers. The Harry S. Truman Library. Independence, Missouri.
Redding, John M. Papers. The Harry S. Truman Library. Independence, Missouri.
Roosevelt, Franklin D. Papers. The Franklin D. Roosevelt Library. Hyde Park, New York.
Rosenman, Samuel I. Papers. The Franklin D. Roosevelt Library. Hyde Park, New York.
Truman, Harry S. Papers. The Harry S. Truman Library. Independence, Missouri.
Webb, James E. Papers. The Harry S. Truman Library. Independence, Missouri.
Yates, Sidney R. Papers. The Harry S. Truman Library. Independence, Missouri.

UNPUBLISHED UNITED STATES GOVERNMENT DOCUMENTS

Department of the Army. Records of the Office of the Chief of Transportation. Federal Records Center. Suitland, Maryland.
Department of Commerce. Records of the Bureau of Public Roads. Record Group 30, Federal Records Center. Suitland, Maryland.
Department of Commerce. Records of the Office of the Secretary. Record Group 40, Department of Commerce. Washington, D.C.
Department of Commerce. Records of the Office of the Under Secretary for Transportation. Record Group 40, Department of Commerce. Washington, D.C.
Housing and Home Finance Agency. Records of the Office of the Administrator. Record Group 207, National Archives. Washington, D.C.
National Resources Planning Board. Records of the National Resources Planning Board. Record Group 187, National Archives. Washington, D.C.
Treasury Department. Records of the Office of the Secretary. Record Group 56, National Archives. Washington, D.C.
United States Senate. Records of the United States Senate. Record Group 46, National Archives. Washington, D.C.

PUBLISHED CONGRESSIONAL DOCUMENTS (LISTED CHRONOLOGICALLY)

Congressional Record. 75th Congress through 84th Congress. 1940–1956.
House of Representatives, Committee on Roads. *Hearings on H.R. 4301: Emergency Construction of Public Highways, Etc.* 74th Congress, 1st Session, 1935.
———. *Hearings on H.R. 7079, H.J. Res. 204 and H.J. Res. 227: Transcontinental Highways, Etc.* 75th Congress, 1st Session, 1937.
Senate, Committee on Post Offices and Post Roads. *Hearings on S. 1771: A*

Bill to Furnish Employment by Providing for Emergency Construction of Public Highways and Related Projects. 75th Congress, 1st Session, 1937.

———. *Hearings on S.J. Res. 106: Transcontinental Highway Commission.* 75th Congress, 1st Session, 1937.

Department of Agriculture, Bureau of Public Roads. *Toll Roads and Free Roads.* House Document No. 272. 76th Congress, 1st Session, 1939.

House of Representatives, Committee on Roads. *Hearings on H.R. 2426: Federal Aid for Post-War Highway Construction,* 2 volumes. 78th Congress, 2d Session, 1944.

———. *Post-War Federal-Aid Highway Act of 1944.* House Report No. 1597 to accompany H.R. 4915. 78th Congress, 2d Session, 1944.

Senate, Special Committee on Post-War Economic Policy and Planning. *The Role of the Federal Government in Highway Development: An Analysis of Needs and Proposals for Post-War Action.* 78th Congress, 2d Session, 1944.

Committee on Post Offices and Post Roads. *Hearings on S. 971, S. 2105.* 78th Congress, 2d Session, 1944.

———. *Post-War Federal-Aid Highway Act of 1944.* Senate Report No. 1056 to accompany S. 2105. 78th Congress, 2d Session, 1944.

Interregional Highways. House Document No. 379. 78th Congress, 2d Session, 1944.

Senate, Subcommittee of the Committee on Public Works, *Hearings on S. 214, S. 1471: Rural Local Roads.* 81st Congress, 1st Session, 1949.

House of Representatives, Committee on Public Works. *Hearings on H.R. 7398, H.R. 7941.* 81st Congress, 2d Session, 1950.

———. *Continuing the Construction of Highways.* House Report No. 1888 to accompany H.R. 7941. 81st Congress, 2d Session, 1950.

Senate, Subcommittee of the Committee on Public Works. *Hearings on S. 3424, H.R. 7941: Federal Aid Highway Act of 1950.* 81st Congress, 2d Session, 1950.

———. Committee on Public Works. *Continuing the Construction of Highways.* Senate Report No. 2044 to accompany H.R. 7941. 81st Congress, 2d Session, 1950.

House of Representatives, Committee on Public Works, Subcommittee on Roads. *Hearings, National Highway Study,* 2 Parts. 83rd Congress, 1st Session, 1953.

———, Committee on Public Works. *Hearings on H.R. 7678, H.R. 7818, H.R. 7841, H.R. 7124, H.R. 7207, H.R. 14, H.R. 1407, H.R. 3528, and H.R. 3529: Federal-Aid Highway Act of 1954.* 83rd Congress, 2d Session, 1954.

———. *Federal-Aid Highway Act of 1954.* House Report No. 1308 to accompany H.R. 8127. 83rd Congress, 2d Session, 1954.

Senate, Committee on Public Works, Subcommittee on Roads. *Hearings on S. 2859, S. 2982, S. 3069, S. 3184: Federal-Aid Highway Act of 1954.* 83rd Congress, 2d Session, 1954.

———, Committee on Public Works. *Federal-Aid Highway Act of 1954.* Senate Report No. 1093 to accompany S. 3184. 83rd Congress, 2d Session, 1954.

House of Representatives, Committee on Public Works. *Hearings on H.R. 4260: National Highway Program,* 2 Parts. 84th Congress, 1st Session, 1955.

———. *National System of Interstate and Defense Highways Act of 1955.* House Report No. 1336 to accompany H.R. 7474. 84th Congress, 1st Session, 1955.

Senate, Subcommittee of the Committee on Public Works. *Hearings on S. 1048, S. 1072, S. 1160, S. 1573: National Highway Program.* 84th Congress, 1st Session, 1955.

———, Committee on Public Works. *Federal-Aid Highway Act of 1955.* Senate Report No. 350 to accompany S. 1048. 84th Congress, 1st Session, 1955.

House of Representatives, Committee on Public Works. *Hearings on H.R. 8836: National Highway Program, Federal-Aid Highway Act of 1956.* 84th Congress, 2d Session, 1956.

OTHER PUBLISHED UNITED STATES GOVERNMENT DOCUMENTS

Department of Commerce, Bureau of Public Roads. *Annual Report.* Washington, D.C., 1950–1956.

———. *Development of the Interstate Highway System.* Washington, D.C., 1964.

———. *Highway Statistics: 1955.* Washington, D.C., 1957.

———. *Highway Statistics: Summary to 1955.* Washington, D.C., 1957.

Department of Transportation. *Highway Statistics: Summary to 1965.* Washington, D.C., 1967.

[Executive Office of the President]. *A Report on the Washington Conference of Governors.* Washington, D.C., 1954.

Federal Works Agency, Public Roads Administration. *Highway Statistics, 1946.* Washington, D.C., 1947.

———. *Highway Statistics: Summary to 1945.* Washington, D.C., 1947.

———. *Work of the Public Roads Administration.* Washington, D.C., 1946–1949.

National Resources Committee. *Our Cities: Their Role in the National Economy.* Washington, D.C., 1937.

National Resources Planning Board. *Transportation and National Policy.* Washington, D.C., 1942.

The President's Advisory Committee on a National Highway Program. *A Ten-Year National Highway Program: A Report to the President.* N.p., 1955.

President's Commission on Intergovernmental Relations, Study Committee on Federal Aid to Highways. *A Study Committee Report on Federal Aid to Highways.* N.p., 1955.

Statutes at Large, 1944–1956.

Study Committee on Federal Aid to Highways. *A Study Committee Report on Federal Aid to Highways.* N.p., 1955.

PROCEEDINGS AND PUBLICATIONS OF BUSINESS AND PROFESSIONAL ORGANIZATIONS AND OF STATE AND LOCAL AGENCIES

American Association of State Highway Officials. "Highway Definitions." *Pro-*

ceedings of the American Association of State Highway Officials, 1949. Washington, D.C., n.d.

American Institute of Planners, Committee on Urban Transportation. *Urban Freeways.* New York, 1947.

American Road Builders' Association, Committee on Elevated Highways. "Report of the Committee on Elevated Highways." *Proceedings of the Thirty-sixth Annual Convention,* 1939. Washington, D.C., n.d.

———. "Report of the Committee on Elevated Highways." *Proceedings of the Thirty-eighth Annual Convention,* 1941. Washington, D.C., n.d.

———, Committee on Federal and State Road Aid to Municipalities. "Report of the Committee on Federal and State Road Aid to Municipalities." *Proceedings of the Thirty-seventh Annual Convention,* 1940. Washington, D.C., n.d.

American Society of Planning Officials, Committee on Highways and Transportation. "Report of the Committee on Highways and Transportation." *Proceedings of the National Conference on Planning,* 1940. Chicago, 1940.

American Trucking Associations, Incorporated. *Statement of Highway Policy.* Washington, D.C., 1951.

Association of Highway Officials of the North Atlantic States. "Resolution." *Proceedings of the Fourteenth Annual Convention of the Association of Highway Officials of the North Atlantic States,* 1938. Trenton, New Jersey, 1938.

Automotive Safety Foundation. *An Engineering Study of Ohio's Highways, Roads and Streets: A Report to the Ohio Program Commission and the Highway Study Committee.* Columbus, 1950.

Chamber of Commerce of the United States. *Policy Declarations of the Chamber of Commerce of the United States.* Washington, D.C., 1949.

———, Construction and Civic Development Department. *City Planning and Urban Development.* Washington, D.C., 1952.

———, Transportation and Communication Department. *Proceedings of the Thirty-seventh Annual Meeting,* 1949. Washington, D.C., n.d.

Cincinnati, City Planning Commission. *The Cincinnati Metropolitan Master Plan and the Official City Plan of the City of Cincinnati.* Cincinnati, 1948.

———. *The Official City Plan of Cincinnati, Ohio.* Cincinnati, 1925.

———. *Program of the Capital Improvements of the City of Cincinnati, 1951–1955.* Cincinnati, 1950.

Kansas City, Missouri, City Planning Commission. *City Planning: Kansas City, Missouri, 1920–1941.* Kansas City, n.d.

———. *The Master Plan for Kansas City.* Kansas City, 1947.

National Highway Users Conference. *Dedication of Special Highway Revenues to Highway Purposes: An Analysis of the Desirability of Protecting Highway Revenues through Amendments to State Constitutions.* Washington, D.C., 1941.

———. *The Eastman Report Finds That Highway Users Pay Their Way and More.* Washington, D.C., 1940.

———. *Highway Taxation, Finance and Administration: An Outline of Policies.* Washington, D.C., 1938.

———. *Highway Transportation Remakes America.* Washington, D.C., n.d.

———. Question and answer session in the *Proceedings of the Second Highway Transportation Congress.* Washington, D.C., 1948.

The Ohio State University. Group discussion meetings in the *Proceedings of the Ohio Highway Engineering Conference.* Engineering Bulletin No. 141. Columbus, 1950.

———. Group Discussion meetings in the *Proceedings of the Ohio Highway Engineering Conference.* Engineering Bulletin No. 145. Columbus, 1951.

State of Connecticut. *Biennial Report of the Highway Commissioner to the Governor for the Fiscal Years Ended June 30, 1943, and June 30, 1944.* Public Document No. 36. N.p., 1944.

State of Michigan. *Twentieth Biennial Report of the State Highway Commissioner for the Fiscal Years Ending June 30, 1943, and June 30, 1944.*

AUTOBIOGRAPHIES, LETTERS, MEMOIRS, AND PERSONAL ACCOUNTS

Adams, Sherman. *Firsthand Report: The Story of the Eisenhower Administration.* New York, 1961.

Benson, Ezra T. *Cross Fire: The Eight Years with Eisenhower.* New York, 1962.

Bettman, Alfred. *City and Regional Planning Papers,* edited by Arthur C. Comey. Cambridge, 1946.

Branyan, Robert L., and Larsen, Lawrence H. (eds.). *The Eisenhower Administration, 1953–1961: A Documentary History.* Volume 1. New York, 1971.

Eisenhower, Dwight D. *Mandate for Change: The White House Years, 1953–1956.* New York, 1963.

Hughes, Emmet J. *Ordeal of Power: A Political Memoir of the Eisenhower Years.* New York, 1963.

Larson, Arthur. *Eisenhower: The President Nobody Knew.* New York, 1968.

———. *A Republican Looks at His Party.* New York, 1956.

Martin, Joseph W. *My First Fifty Years in Politics.* New York, 1960.

Public Papers of the Presidents of the United States: Dwight D. Eisenhower, 1953–1956. Washington, D.C., 1958–1960.

Public Papers of the Presidents of the United States: Harry S. Truman, 1946–1953. Washington, D.C., 1962–1966.

Rosenman, Samuel I. (ed.). *The Public Papers and Addresses of Franklin D. Roosevelt.* Volumes 7–13. New York, 1941, 1950.

Truman, Harry S. *Memoirs: Years of Decisions.* Volume I. New York, 1955.

———. *Memoirs: Years of Trial and Hope.* Volume II. New York, 1956.

PERIODICALS

American Highways, 1949–1956.

Highway Highlights: Automotive Transportation in All Its Phases, 1942–1956.

New York Times, 1939, 1955.

Public Roads: A Journal of Highway Research, 1938–1956.

CONTEMPORARY ARTICLES AND BOOKS

Abel, Edson. "The Case against the Collier Highway Program." *Transactions of the Commonwealth Club of California* 91 (1947), 133–136.

Abelard, A. R. "Benefit-Cost Ratio Method of Computing Priorities of Construction." University of Colorado. *Proceedings of the Twenty-second Annual Highway Conference.* Highway Series No. 22 (July, 1949), 46–49.

Adams, Thomas. *Planning the New York Region: An Outline of the Organization, Scope and Progress of the Regional Plan.* New York, 1927.

Alexander, Harry W. "Bringing Interregional Thoroughfares to the Hearts of the Cities: Federal-Aid Highway Act Offers Unique Opportunity for Improving Conditions." *The American City* 60 (February, 1945), 59–60, 93.

"ATA Convention Launches Public Relations Program." *Power Wagon: The Motor Truck Journal* (November, 1939), 12–14.

"ATA Convention Pinpoints Highway Problems." *Commercial Car Journal: The Magazine for Fleet Operators with Which Is Combined Operation and Maintenance* (December, 1951), 64–66, 150, 152, 154, 157.

"ATA Convention Sights for 1955." *Power Wagon: The Motor Truck Journal* (November, 1954), 12–15.

"The Automobiles Are Coming." *The American City* 58 (May, 1943), 85, 87.

Bagby, Scott. "Protecting Good Neighborhoods from Through-Traffic Decline." *Traffic Quarterly* 8 (October, 1954), 410–422.

Baker, Donald M. "Financing Express Highways in Metropolitan Areas." *The American City* 61 (October, 1946), 93–94.

Balfour, Frank C. "Acquisition of Access Rights in California." Highway Research Board, *Proceedings of the Twenty-fourth Annual Meeting* (n.d), 16–23.

———. "Effect of Freeway Development on Adjacent Land Values in California." *Proceedings of the Thirty-third Annual Meeting of the American Association of State Highway Officials* (no imprint), 55–76.

Barnett, Joseph. "Express Highway Planning in Metropolitan Areas." *Proceedings of the American Society of Civil Engineers* 72 (March, 1946), 287–305.

———. "Expressways in Relieving Urban Traffic Congestion." *Traffic Engineering* 23 (February, 1953), 168–169.

———. "The Highway in Urban and Suburban Areas," in Labatut, Jean, and Lane, Wheaton J. (eds). *Highways in Our National Life: A Symposium.* Princeton, New Jersey, 1950.

———. "Intersection Design for Arterial Routes in Urban Areas." University of Michigan. *Proceedings of the Thirty-second Annual Highway Conference* 49 (1947), 25–38.

———. "Urban and Inter-City Road Improvement." The Ohio State University. *Proceedings of the Ohio Highway Engineering Conference.* Bulletin No. 129 (September, 1947), 25–34.

Bartholomew, Harland. *Development and Planning of American Cities: An Address before the Student Body of the Carnegie Institute of Technology.* Carnegie Press Occasional Paper No. 1, Pittsburgh, 1950.

———. "Effect of Urban Decentralization Upon Transit Operations and Policies." *Proceedings of the Fifty-ninth Annual Convention of the American Transit Association and Its Affiliated Organizations* (1940), 481–488.

———. "The Location of Interstate Highways in Cities." *American Planning and Civic Annual* (1949), 73–78.

———. "The Neighborhood—Key to Urban Redemption." *American Planning and Civic Annual* (1941), 243–246.

———. "Planning for Metropolitan Transportation." *Planning and Civic Comment* 18 (September, 1952), 1–4.

"Basic Truck Operating Problems Get Airing at ATA Spring Meeting." *Power Wagon: The Motor Truck Journal* (June, 1949), 18–19.

Beach, D. M. "The United States Bureau of Public Roads." *Roads and Streets* 82 (May, 1939), 48–56.

Bean, E. Carroll. "Resolved, 'That the Toll Method of Financing Highways Is Unsound.'" *Proceedings of the Third Highway Transportation Congress* (1950), 49–50.

Bellis, Wesley R. "Costs of Traffic Inefficiencies." *Traffic Engineering* 23 (November, 1952), 53–55, 60.

Bevis, Howard W. "The Application of Benefit-Cost Ratios to an Expressway System." Highway Research Board, *Proceedings of the Thirty-fifth Annual Meeting* (1956), 63–75.

Blucher, Walter H. "Planning for the Post-War Period." Purdue University. *Proceedings of the Twenty-eighth Annual Road School* 26 (March, 1942), 30–39.

Boykin, L. E. "Interregional Highways: Legal and Right-of-Way Problems." *Roads and Streets* 83 (January, 1940), 57–60.

Brandt, A. W. "Shaping Our Highway Program for National Defense." *American Highways* 19 (October, 1940), 15–19.

Bresnahan, William A. "Freight Transportation on the Highway," pp. 247–254 in Labatut, Jean, and Lane, Wheaton J. (eds.). *Highways in Our National Life: A Symposium.* Princeton, New Jersey, 1950.

———. "Truck Transportation—From the Trucker's Viewpoint." The Ohio State University. *Proceedings of the Ohio Highway Engineering Conference.* Bulletin No. 145 (September, 1951), 10–16.

———. "Who Should Pay How Much of Highway Costs." *Commercial Car Journal: The Magazine for Fleet Operators with Which Is Combined Operation and Maintenance* (July, 1952), 51–53, 174, 176, 178, 180, 182, 222, 224.

Brownlee, O. H., and Heller, Walter W. "Highway Development and Financing." *The American Economic Review* 96 (May, 1956), 232–264.

Buckley, J. P. "Ohio Takes Inventory of Its Roads and Streets." The Ohio State University. *Proceedings of the Ohio Highway Engineering Conference.* Bulletin No. 136 (March, 1949), 21–32.

Buckley, J. P., and Fritts, Carl E. "Objectives and Findings of Highway Needs Studies." Highway Research Board, *Proceedings of the Twenty-eighth Annual Meeting* (1949), 1–13.

Butler, Arthur C. "Adding Up the Legislative Year." *Commercial Car Journal: The Magazine for Fleet Operators with Which Is Combined Operation and Maintenance* (August, 1953), 80–81, 170, 172.

———. "Legislative Outlook for 1949." *Commercial Car Journal: The Magazine for Fleet Operators with Which Is Combined Operation and Maintenance* (November, 1948), 98–102, 175–176, 178, 180, 182.

———. "1955 Highway User Legislative Roundup." *Commercial Car Journal:*

The Magazine for Fleet Operators with Which Is Combined Operation and Maintenance (September, 1955), 80–83, 114–116, 119.

———. "Road Blocks to Highway Progress." Purdue University. *Proceedings of the Thirty-fifth Annual Road School* 33 (September, 1949), 39–45.

Buttenheim, Harold S. "City Highways and City Parking—An American Crisis." *The American City* 61 (November, 1946), 116–117, 123, 139.

Cabell, Henry F. "The Economic Aspects of Interregional Highways." *Roads and Streets* 83 (January, 1940), 61–62.

Cartwright, Wilburn. "The Congressional Outlook." *Proceedings of the Thirty-eighth Annual Meeting of the American Road Builders' Association* (1941), 32–35.

Chamber of Commerce of the United States, Transportation and Communication Department. "Merchandise Pickup and Delivery." *Power Wagon: The Motor Truck Journal* 85 (August, 1950), 9–10, 12–13; (September, 1950), 12, 14, 16; (October, 1950), 28, 30.

Chaney, R. C. "Cleveland Freeways a Major Problem: Metropolitan Cleveland Selected as a Major Objective in Reconditioning the State's System by the Ohio Department of Highways under Director Hal G. Sours." *Roads and Streets* 86 (June, 1943), 40–44.

Churchill, Henry S. "Planning in a Free Society." *Journal of the American Institute of Planners* 20 (Fall, 1954), 189–191.

Claire, William H. "Redevelopment and Traffic." *Traffic Engineering* 24 (November, 1953), 54–56.

Clarkeson, John. "Urban Expressway Location." *Traffic Quarterly* 7 (April, 1953), 252–260.

"Cleveland: City with a Deadline." *Architectural Forum* 103 (August, 1955), 130–139.

Cobo, Albert. "Street and Traffic Management: A New Approach." *Traffic Quarterly* 6 (January, 1952), 40–50.

Cookingham, L. P. Comments to a Panel on Expressways and the Central Business District. *American Planning and Civic Annual* (1954), 140–146.

Coons, H. C. "Detroit Pushes Expressway System through Congested Urban Area." *Civil Engineering* 24 (January, 1954), 18–21.

Cope, Edwin, and Meadows, Richard W. "Road-User and Property Taxes on Selected Motor Vehicles." Highway Research Board, *Proceedings of the Thirty-second Annual Meeting* (1953), 12–44.

Cox, M. E. "A Highway and Transportation Plan Emerges." *Civil Engineering* 24 (March, 1954), 141–143.

Crane, Warren E. "Highway Transport, the Imprisoned Giant." *Power Wagon: The Motor Truck Journal* 7 (August, 1947), 14 ff.

Cronk, Duane L. "1954: Turning Point Year." *Roads and Streets* 98 (January, 1955), 44–45.

Crowson, J. K. "The Need for a Decision on Future Federal-Aid Policy." *Proceedings of the Forty-first Annual Meeting of the American Association of State Highway Officials* (c. 1956), 27–30.

Curtiss, C. D. "The Highway Situation in Washington." *Proceedings of the Fourteenth Annual Convention of the Association of Highway Officials of the North Atlantic States* (1938), 12–17.

Davis, Harmer E. "Supplementary Comments on Credit Financing for Highways." *Proceedings of the Thirty-eighth Annual Meeting of the American Association of State Highway Officials* (n.d.), 125–129.

———, et al. "Some Recent Aspects of the Toll-Road Situation." *Proceedings of the Thirty-eighth Annual Meeting of the American Association of State Highway Officials* (n.d.), 95–124.

Davis, Harold L. "How Post-War Planning Is Organized in Toledo." *The American City* 59 (May, 1944), 58.

Dearing, Charles L. "Toll Road Rates and Highway Pricing." *The American Economic Review* 97 (May, 1957), 441–452.

———. *American Highway Policy*. Washington, D.C., 1941.

———, and Owen, Wilfred. *National Transportation Policy*. Washington, D.C., 1949.

Delano, G. H. "Super-Highways and Primary Roads." *Proceedings of the Fourteenth Annual Convention of the Association of Highway Officials of the North Atlantic States* (1938), 69–74.

Demarsh, Lee F. "John F. Fitzgerald Expressway Boston Central Artery." *Traffic Quarterly* 10 (October, 1956), 454–470.

"Detroit Expressway System Designed by Engineer W. Earl Andrews Eliminates Downtown Congestion, Speeds up Traffic Movement Between Widely Separated Districts." *The Architectural Forum* (September, 1945), 125–129.

Downs, E. F. "What's Ahead for Highways and Transportation." *Public Utilities Fortnightly* 26 (September, 1940), 396–405.

Dudley, Arthur S. "Sacramento's Post-War Plans Two Major Fronts: Public Works and Private Industry." *The American City* 58 (October, 1943), 45–46.

Duzan, Hugo C. "Vehicular Charges on Highway Toll Facilities." Highway Research Board, *Proceedings of the Thirty-second Annual Meeting* (1953), 45–67.

———, et al. "Recent Trends in Highway Bond Financing." Highway Research Board, *Proceedings of the Thirty-first Annual Meeting* (1952), 1–25.

Eckhardt, Harold. "Traffic Bottlenecks in Cities." The Ohio State University. *Proceedings of the Ohio Highway Engineering Conference*, Engineering Bulletin No. 136 (March, 1949), 216–219.

Elder, Herbert W. "Houston's Urban Expressways." *Traffic Quarterly* 3 (April, 1949), 166–173.

Ellis, James G. "Thumbs Down on Toll Roads." *Commercial Car Journal: The Magazine for Fleet Operators with Which Is Combined Operation and Maintenance* (June, 1939), 30 ff.

Emery, George F. "Urban Expressways." *American Planning and Civic Annual* (1947), 123–129.

Erickson, E. Gordon. "The Superhighway and City Planning: Some Ecological Considerations with Reference to Los Angeles." *Social Forces* 28 (May, 1950), 429–434.

Evans, Henry K. "Can We Afford Model T Roads?" *Commercial Car Journal: The Magazine for Fleet Operators with Which Is Combined Operation and Maintenance* (August, 1952), 51, 114, 116, 119–120.

———. "The Great Highway Robbery—Or Is It?" *Commercial Car Journal: The Magazine for Fleet Operators with Which Is Combined Operation and Maintenance* (September, 1953), 64–65, 106, 108, 110, 112.

———. "Looking Ahead in Urban Transportation." *Traffic Engineering* 21 (March, 1951), 189–192.

———. "Parking and Its Importance to the Downtown Business District." *Traffic Engineering* 24 (December, 1953), 89, 91–95.

Fairbank, H[erbert] S. "Interregional Highways Indicated by State-Wide Highway Planning Surveys." *Roads and Streets* 83 (January, 1940), 37–44.

———. "The Interstate Highway System." *Proceedings of the Second Highway Transportation Congress* (circa 1948), 12–13.

———. "Military Highways." University of Michigan. *Proceedings of the Twenty-seventh Annual Highway Conference* 43 (July, 1941), 35–43.

———. "A New Highway Program." *American Planning and Civic Annual* (1944), 44–51.

———. "Planned Highway Transportation." University of Colorado. *Proceedings of the Twentieth Annual Highway Conference.* Highway Series No. 20 (April, 1947), 3–10.

Farrell, Fred B., and Paterick, Henry R. "The Capital Investment in Highways." Highway Research Board, *Proceedings of the Thirty-second Annual Meeting* (1953), 1–11.

"First Underpass in Washington, D.C., Will Have Carbon Monoxide Detector and Ventilating System." *Roads and Streets* 82 (July, 1939), 21–24.

Fisk, Chester C. "Modernizing California's Highways." *Transactions of the Commonwealth Club of California* 91 (1947), 87–132.

Flanakin, H. A. "Major Traffic Problems as They Appear to the Trucking Industry." *Proceedings of the Tennessee Highway Conference.* The University of Tennessee *Record* 57 (July, 1954), 67–70.

Ford, Stanley H. "The Military Requirement of Our Highway System." *Proceedings of the Thirty-seventh Annual Convention of the American Road Builders' Association* (1940), 68–71.

Forest, Ernest S. "Reciprocity Crisis . . . Threat to Profitable Truck Use." *Commercial Car Journal: The Magazine for Fleet Operators with Which Is Combined Operation and Maintenance* (March, 1954), 64–65, 128, 133–134, 136.

Foster, W. S. "The Duquesne Way." *The American City* 59 (January, 1944), 40–42.

Fritts, C. E. "California's Highway Needs—2." University of California, The Institute of Transportation and Traffic Engineering. *Proceedings of the Fifth California Street and Highway Conference* (1953), 13–21.

Gardner, Lamar W. "The Economy of Freeways." *Civil Engineering* 24 (December, 1953), 83–86, 96.

Gibbons, J. W. "The Economic Costs of Traffic Congestion." *Proceedings of the Tennessee Highway Conference.* University of Tennessee *Record* 57 (July, 1954), 6–14.

Goodrich, E[rnest] P., and Ford, George B. *Report of Suggested Plan of Procedure for City Planning Commission, City of Jersey City, New Jersey.* N.p., 1913.

Graves, Richard. "California's Urban Transportation Problem." University of California, The Institute of Transportation and Traffic Engineering. *Proceedings of the Fifth California Street and Highway Conference* (1953), 83–85.

Gray, Chester H. *Transportation in 1950.* Washington, D.C., c. 1940.

Gray, H. Gordon. "Cheaper by the Mile." *Proceedings of the American Association of State Highway Officials* (n.d.), 19–27.

Greer, DeWitt C. "Balancing the Highway Needs for Both Rural and Urban Areas." *Traffic Quarterly* 6 (July, 1952), 331–338.

Greer, M. V. "State Assistance in the Development of Freeways and Major Arterial Routes in Urban Areas in Texas." *Traffic Engineering* 23 (March, 1953), 198–199.

Gregg, Ronald E. "Toledo and Its Big Tomorrow." *National Municipal Review* 34 (November, 1945), 493–498.

Grumm, Fred J. "California's Plan for Freeways in Metropolitan Areas." *Civil Engineering* 2 (October, 1941), 569–572.

Hadley, Homer M., et al. "Express Highway Planning in Metropolitan Areas." *Proceedings of the American Society of Civil Engineers* 72 (September, 1946), 1057–1069.

Haldeman, B. Antrim. "The Street Layout." *The Annals* 51 (January, 1914), 182–191.

Hale, Hal H. "Highways and Highway Finance." *Proceedings of the Tennessee Highway Conference.* University of Tennessee *Record* 57 (July, 1954), 71–74.

Hammond, Harold F. "What Percent of Highway Construction Costs Should Be Paid by the Federal Government." *Traffic Engineering* 20 (March, 1950), 229–230.

Harper, Robert E. "Roads for Defense." *Florida Public Works* 17 (August, 1940), 9–10, 16.

Hart, William D. "The Use of Federal-Aid Funds in Bond Retirement." *Traffic Engineering* 25 (February, 1955), 191–192.

Harter, Fred. "The Ohio Highway Study Committee, Its Function and Purpose." The Ohio State University. *Proceedings of the Ohio Highway Engineering Conference.* Bulletin No. 141 (May, 1950), 37–40.

Henderson, Robert S. "Freeway System of Benefits." *Traffic Engineering* 26 (March, 1956), 248–250, 256.

Hewes, Lawrence I. "Metropolitan Freeways and Mass Transportation." *Transactions of the Commonwealth Club of California* 90 (August, 1946), 99–131.

Hewitt, Arthur F. "Freeways." The University of Colorado. *Proceedings of the Colorado Highway Conference* (1942), 21–24.

"Highways and Horizons Depicts Desirable Highway System for 1960." *Roads and Streets* 82 (July, 1939), 62, 64.

Hilts, H. E. "The 1946 Federal-Aid Highway Program." University of Michigan. *Proceedings of the Thirty-first Annual Highway Conference* 48 (July, 1946), 132–137.

Hirsch, Phil. "Chicago Fleetman Battle Traffic Bottlenecks." *Commercial Car Journal: The Magazine for Fleet Operators with Which Is Combined Operation and Maintenance* (June, 1954), 68–71, 178, 180, 182.

Hoffman, Paul G. "Interregional Highways: America's Highways—New Frontiers in National Progress." *Roads and Streets* 83 (January, 1940), 35–36.

"The Houston Expressway." *The American City* 63 (November, 1948), 116–117.

Howard, Randall R. "Coast-to-Coast Trucking Shows Steady Upward Trend." *Power Wagon: The Motor Truck Journal* 83 (December, 1949), 9–11.

Hoyt, Homer. "The Influence of Highways and Transportation on the Structure and Growth of Cities and Urban Land Values," pp. 201–206 in Labatut, Jean, and Lane, Wheaton J. (eds.). *Highways in Our National Life: A Symposium.* Princeton, New Jersey, 1950.

Hudgins, B. "Denver–Boulder Turnpike." The University of Colorado. *Proceedings of the Twenty-fourth Annual Highway Conference.* Highway Series No. 24 (May, 1951), 18–20.

Johnson, Pyke. "Highway Post-War Planning." Purdue University. *Proceedings of the Thirty-first Annual Road School* 29 (March, 1945), 29–31.

———. "Highway Transportation in the Defense Effort." *Proceedings of the Thirty-eighth Annual Convention of the American Road Builders' Association* (1941), 28–31.

———. "Highways for Post-War Prosperity." *Commercial Car Journal: The Magazine for Fleet Operators with Which Is Combined Operation and Maintenance* (March, 1944), 54–55, 196, 198, 200.

Jones, Bethune. "Automotive Taxes." *American Cattle Producer* 27 (December, 1945), 23–24.

Jorgensen, Roy E. "Better Roads with PAR." The Ohio State University. *Proceedings of the Ohio Highway Conference.* Engineering Series Bulletin No. 150 (November, 1952), 62–72.

———. "Financing a Nation-Wide Highway Program." Highway Research Board, *Proceedings of the Twenty-ninth Annual Meeting* (1950), 1–19.

———. "Financing the Highway Program." *Proceedings of the Thirty-fourth Annual Convention of the American Association of State Highway Officials* (1948), 57–63.

———. "The Highway Users Interest in Highway Reports." *Proceedings of the Fortieth Annual Convention of the American Association of State Highway Officials* (n.d.), 76–80.

———. "Sizes of State Highway Programs: How Big Should They Be?" *Traffic Quarterly* 6 (January, 1952), 59–67.

Kaltenbach, Henry J. "Proposed Federal Legislation for Highways." *Proceedings of the Forty-first Annual Meeting of the American Association of State Highway Officials* (c. 1956), 47–55.

Kauer, T. J. "Ohio's First Cloverleaf." *Roads and Streets* 83 (April, 1940), 35–39.

———. "Ohio Turnpike Project No. 1." *Traffic Quarterly* 7 (January, 1953), 108–123.

Kelcey, Guy. "Problems of Freeway Development." *Proceedings of the Thirty-sixth Annual Meeting of the American Association of State Highway Officials* (1950), 362–369.

Keller, W. J. "Highways of the Future." The University of Colorado. *Papers Presented at the Highway Conference* (1942), 4–11.

Kelly, James F. "Development and Progress of Superhighways in Cook County, Illinois." Purdue University. *Proceedings of the Thirty-fifth Annual Road School.* Extension Series No. 69 (September, 1949), 246–258.

Kennedy, G. Donald. "Engineering a Highway Program for Tomorrow." The Ohio State University. *Proceedings of the Ohio Highway Engineering Conference.* Experiment Station Bulletin No. 134 (November, 1948), 6–20.

———. "Implications of Recent Highway Planning." *Proceedings of Confer-*

ence on Highways of the Future. The University of Tennessee *Record* 50 (July, 1947), 73–79.

———. "Long-Range Highway Planning." *Proceedings of the Second Highway Transportation Congress* (1948), 11–12.

———. "The Planning of the Highway," pp. 290–298 in Labatut, Jean, and Lane, Wheaton J. (eds.). *Highways in Our National Life: A Symposium.* Princeton, New Jersey, 1950.

———. "Planning the Statewide System of Highway Transportation." *Traffic Quarterly* 3 (July, 1949), 201–213.

———. "Will Our Cities Survive." *Traffic Engineering* 22 (August, 1952), 405–406.

Kenney, James B. "Problems of the Western Contractor." *Proceedings of the Thirty-sixth Annual Convention of the American Road Builders' Association* (1939), 445–453.

Kent, T[heodore] J. "City and Regional Planning Needs in Relation to Transportation." The University of California, Institute of Transportation and Traffic Engineering. *Proceedings of the First California Institute on Street and Highway Problems.* Technical Bulletin No. 1 (1949), 55–58.

Kingery, Robert. "Interregional Highways from the Municipal Viewpoint." *Roads and Streets* 83 (January, 1940), 45–48.

———. "Regional Plan of Chicago." *Proceedings of the Thirty-seventh Annual Convention of the American Road Builders' Association* (c. 1940), 503–511.

Kitfield, P. H. "The Future of Highway Building in New England." *Proceedings of the Thirty-eighth Annual Convention of the American Road Builders' Association* (c. 1941), 129–131.

Knetzger, Austin C. "Fitting Truck Movement Efficiently into the Traffic Problem." *Traffic Quarterly* 5 (July, 1951), 312–324.

La Guardia, F[iorello] H. "Sodium-Lighted Shoestring Park Encircling New York." *The American City* 55 (August, 1940), 103, 105, 107.

"Lakefront Freeway Speeds Cleveland Traffic." *Engineering News-Record* 124 (April 11, 1940), 64–67.

Lauderbaugh, C. P. "Franklin County, Ohio Plans." *American Planning and Civic Annual* (1953), 36–37.

Lautzenheiser, Fred B. "Effect of the AASHO Code on Truck Design." *Commercial Car Journal: The Magazine for Fleet Operators with Which Is Combined Operation and Maintenance* (July, 1946), 38–42.

Levin, David R. "Financing Highway Right-of-Way." *Proceedings of the Thirty-fifth Annual Meeting of the American Association of State Highway Officials* (1949), 94–117.

———. "The Highway and Land Use," pp. 268–276 in Labatut, Jean, and Lane, Wheaton J. (eds.). *Highways in Our National Life: A Symposium.* Princeton, New Jersey, 1950.

———. "Legislative and Administrative Implementation of the Post-War Highway Program." Highway Research Board, *Proceedings of the Twenty-fourth Annual Meeting* (Unassembled). (N.d.), 4–15.

———. "Limited Access Highways in Urban Areas." *The American City* 59 (February, 1944), 77, 79, 81.

———. "Needed: Better Enabling Laws on Parking Facilities." *The American City* 62 (February, 1947), 113.

———. "Parking Facilities as Public Utilities." Highway Research Board, *Proceedings of the Thirtieth Annual Meeting* 30 (1951), 15–24.

———. "State Highway Department Authority in Municipalities." *Proceedings of the Thirty-fifth Annual Convention of the American Association of State Highway Officials* (1949), 80–88.

———, et al. Panel on Expressways and the Central Business District. *American Planning and Civic Annual* (1954), 131–139.

Lewarch, E. E. "Seattle's Freeway System." *Traffic Quarterly* 8 (July, 1954), 300–320.

Lewis, Harold M., and Goodrich, Ernest P. *Highway Traffic.* Vol. III. New York, 1927.

Livingston, Robert E. "The Sufficiency Survey Method of Determing [sic] Priorities for Construction Methods." The University of Colorado. *Proceedings of the Twenty-second Annual Highway Conference.* Highway Series No. 22 (July, 1949), 41–45.

Lochner, Harry W. "Express Highways in the Chicago Metropolitan Area." Purdue University. *Proceedings of the Twenty-seventh Annual Road School.* Extension Series No. 50 (May, 1941), 101–112.

———. "Express Highway Planning in Chicago Metropolitan Area." *Proceedings of the Highway Conference.* The University of Tennessee *Record* 47 (July, 1944), 48–59.

———. "Express Highway Planning in Metropolitan Areas." *Proceedings of the American Society of Civil Engineers* 72 (May, 1946), 722–724.

"Long Distance Trucking Handicapped by Highway Shortcomings: Bottleneck Highways and Inadequate Bridges Slow up Intercity Trucking and Add to Nation's Haulage Bill." *Power Wagon: The Motor Truck Journal* 77 (July, 1946), 9–11.

McCallum, William R. *Highway Bond Financing . . . An Analysis.* Washington, D.C., 1963.

MacCleery, Russell E. "The Place of the Motor Vehicle in Post-War Transportation." *Proceedings of the Twentieth Annual Meeting of the Association of Highway Officials of the North Atlantic States* (Trenton, New Jersey, 1944), 60–77.

McCormack, C. M. "Ohio's Highway Needs." The Ohio State University. *Proceedings of the Ohio Highway Engineering Conference.* Engineering Series Bulletin No. 145 (September, 1951), 42–46.

McCracken, Dwight. "Highway Traffic and Operating Problems Involved in Truck Transportation." *Traffic Engineering* 20 (November, 1949), 74–76.

McDevitt, Frank J. "Facilitating Traffic and Local Transportation." *The American City* 53 (November, 1938), 103, 105.

MacDonald, Thomas H. "The Case for Urban Expressways: Long-Range Planning of Adequate Highway Facilities Will Save Many Cities from Stagnation and Decay." *The American City* 62 (June, 1947), 92–93.

———. "The City's Place in Post-War Highway Planning: Concrete Advice on What Cities Should Do to Help Provide the Economic 'Backlog' against Economic Recession." *The American City* 58 (February, 1943), 42–44.

———. "Highways and National Defense." *American Highways* 19 (October, 1940), 11–14.

———. "The Interstate System in Urban Areas." *American Planning and Civic Annual* (1950), 114–119.

———. "National Road Building Through the Cooperation of the States and the Federal Government." *Roads and Streets* 82 (December, 1939), 50, 52, 54, 56.

———. "The New Federal Highway Program." *American Planning and Civic Annual* (1941), 51–55.

———. "Public Roads and City Planning." *Report of the Urban Planning Conferences Under the Auspices of the Johns Hopkins University.* Baltimore, 1944.

McGavin, C. T. "Traffic and Urban Decentralization." *National Municipal Review* 30 (January, 1941), 28–34.

MacLachlan, K. A. "Engineering and Economic Justification for Major Urban Transportation Improvements and Value of Origin-Destination Surveys." University of California, Institute of Transportation and Traffic Engineering. *Proceedings of the First California Institute of Transportation and Traffic Engineering.* Technical Bulletin No. 1 (1949), 205–212.

McMonagle, J. Carl. "A Comprehensive Method of Scientific Programming." Highway Research Board, *Proceedings of the Thirty-fifth Annual Meeting* (1956), 33–37.

McWhorter, B. P. "Construction Problems Are Mainly Concerned with Public Relations." *Civil Engineering: The Magazine of Engineered Construction* 24 (March, 1954), 147–149.

"The Magic City of Progress." *The American City* 54 (July, 1939), 40–41.

Markham, Baird H. "Elimination of Fiscal Abuses in Highway Administration and Planning." *Proceedings of the Second Highway Transportation Congress* (1948), 17–18.

Martin, Park H. "Pittsburgh's Golden Triangle," *American Planning and Civic Annual* (1951), 138–144.

———. "Urban Highway Development." *Proceedings of the Second Highway Transportation Congress* (1948), 14–15.

Mayer, Harold M. "Moving People and Goods in Tomorrow's Cities." *Annals of the American Academy of Political and Social Science* 242 (November, 1945), 116–128.

Meigs, H. G. "Regulating Highway Construction Practices." *Proceedings of the Thirty-sixth Annual Convention of the American Road Builders' Association* (c. 1939), 414–417.

Merriam, Charles E. "Make No Small Plans." *National Municipal Review* 32 (February, 1943), 63–67.

———. "The National Resources Planning Board: A Chapter in American Planning Experience." *The American Political Science Review* 38 (December, 1944), 1075–1088.

Miller, Fred. "Colorado Limited Access Road: Three Contractors 'Ganged Up' on $1,500,000 Project Which Includes Barrier Curbs and Service Roads at Independent Grades." *Roads and Streets* 86 (October, 1943), 64–68.

Miller, Spencer, Jr. "History of the Modern Highway in the United States," pp. 88–119 in Labatut, Jean, and Lane, Wheaton J. (eds.). *Highways in Our National Life: A Symposium.* Princeton, New Jersey, 1950.

Mitchell, Robert B. "Coordination of Highway and City Planning." Highway

Research Board, *Proceedings of the Twenty-eighth Annual Meeting* (1949), 13–18.

———, and Rapkin, Chester. *Urban Traffic: A Function of Land Use.* New York, 1954.

Moses, Robert M. "The Car and the Road." *Traffic Engineering* 24 (April, 1954), 233–235.

———. "Comprehensive Parkway System of New York Metropolitan Region." *Civil Engineering* 9 (March, 1939), 160–162.

———. "Parks, Parkways, Express Arteries, and Related Plans for New York City after the War." *The American City* 58 (December, 1943), 53–58.

———. "Urban Traffic." *Proceedings of the Thirty-third Annual Convention of the American Association of State Highway Officials* (1947), 150–154.

Mowbray, John McC. "Combatting Urban Blight by Neighborhood Reconstruction." *The American City* 53 (December, 1938), 41–42.

Mullady, Walter. "Highway Haulage Not Subsidized." *Power Wagon: The Motor Truck Journal* 62 (March, 1939), 5–7.

Myers, Fred J. "Interstate Reciprocity in the Regulation and Taxation of Motor Vehicles." Highway Research Board, *Proceedings of the Twenty-seventh Annual Meeting* 27 (c. 1948), 7–10.

Nelson, James C. "Highway Finance Problems in the West." Highway Research Board, *Proceedings of the Thirtieth Annual Meeting* (c. 1951), 9–15.

"New Design Standards for Secondary and Interstate Highways." *Roads and Streets* 88 (November, 1945), 83–85.

Nichols, C. G. "The Merritt Parkway." *Roads and Streets* 83 (March, 1940), 66, 69–70.

"1955: Roadbuilding Up 25%." *Roads and Streets* 98 (January, 1955), 46–47, 51–52, 52H.

Noble, Charles M. "Highway Planning in Metropolitan Areas." *American Planning and Civic Annual* (1948), 104–112.

Nolen, John, et al. *City Plan: Columbus, Georgia.* Cambridge, 1926.

Nolting, Orin F., and Opperman, Paul. *The Parking Problem in Central Business Districts with Special Reference to Off-Street Parking.* Public Administration Service Publication No. 64. Chicago, 1938.

Norton, C. McKim. "Metropolitan Planning." *Traffic Quarterly* 3 (October, 1949), 367–377.

———. "Metropolitan Transportation," pp. 77–91 in Breese, Gerald, and Whitman, Dorothy E. (eds.). *An Approach to Urban Planning.* Princeton, New Jersey, 1953.

"Ohio Gets Set for Its Biggest Road Program." *Engineering News-Record* 144 (May 11, 1950), 32–34.

Ormsbee, D. W. "The North-South Interstate Highway in Colorado." The University of Colorado. *The Twenty-third Annual Highway Conference of the University of Colorado.* Highway Series No. 23 (April, 1950), 50–67.

"Our Short-Changed City Streets." *The American City* 68 (November, 1953), 17.

Owen, Wilfred. *Automobile Transportation in Defense or War: A Report Prepared for the Defense Transport Administration.* Washington, D.C., 1951.

[———]. "The Future Development of Highway Transportation," Highway Research Board, *Proceedings of the Twenty-first Annual Meeting* (c. 1942), 15–22.

———. *The Metropolitan Transportation Problem.* Washington, D.C., 1956.

[———]. "The Problem of Parking Facilities." Highway Research Board, *Proceedings of the Twentieth Annual Meeting* (c. 1941), 27–43.

———. "Problems and Implications of the National Highway Program." *American Highways* (July, 1950), 12–13, 22–23.

———. "Trends in Highway Financial Practices." Highway Research Board, *Proceedings of the Nineteenth Annual Meeting* (c. 1940), 15–49.

———, and Dearing, Charles L. *Toll Roads and the Problem of Highway Modernization.* Washington, D.C., 1951.

Pancoast, D. F. "Ohio Incremental Study: An Experiment in Vehicle-Tax Allocation." Highway Research Board, *Proceedings of the Thirty-second Annual Meeting* 32 (1953), 68–70.

"A Prescription for Saving Downtown Cincinnati." *National Real Estate Journal* 41 (March, 1941), 16–18.

"President's $50 Billion Road Program Stirs Controversies." *Roads and Streets* 97 (August, 1954), 51.

"Private Carriers Defend Rights." *Power Wagon: The Motor Truck Journal* (February, 1955), 15–17, 37.

"'Protected' Highways Urged for Essex County, New Jersey." *The American City* 54 (August, 1939), 79.

"Redevelopment F.O.B. Detroit." *Architectural Forum: The Magazine of Building* 102 (March, 1955), 116–125.

Regional Planning Conference of Los Angeles County. *Proceedings of the First Regional Planning Conference of Los Angeles County, California at Pasadena, January 21, 1922.* No imprint.

Reinhold, Paul B. "A Military Highway System at No Extra Cost." The Ohio State University. *Proceedings of the Ohio Highway Engineering Conference.* Experiment Station Bulletin No. 150 (November, 1952), 7–14.

"Remodeled Street Triples Traffic Room." *Engineering News-Record* 124 (March 28, 1940), 42–46.

"A Report on Roads." *Commercial Car Journal: The Magazine for Fleet Operators with Which Is Combined Operation and Maintenance* (March, 1941), 30, 92, 94, 96.

Rice, Roland. "Roads Were Built for Commerce Not Sightseeing." *Commercial Car Journal: The Magazine for Fleet Operators with Which Is Combined Operation and Maintenance* (November, 1938), 28–29, 114, 116, 118.

Richards, Glenn C. "Does the 1951 Highway Act Meet Today's Needs of Cities?" University of Michigan. *Proceedings of the Thirty-eighth Annual Highway Conference* 55 (c. 1953), 16–18.

Richardson, George S. "A Bold Attack on an Urban Traffic Problem: Penn-Lincoln Parkway." The Ohio State University. *Proceedings of the Ohio Highway Engineering Conference.* Experiment Station Bulletin No. 136 (March, 1949), 55–67.

"Roads For National Defense." *Engineering News-Record* 124 (January 18, 1940), 79–82.

Robinson, Charles M. "The Sociology of a Street Layout." *The Annals* 51 (January, 1914), 192–199.

Rosenberg, Jess N. Remarks to the Commonwealth Club of California. *Transactions of the Commonwealth Club of California* 100 (February, 1956), 74–75.

Ryan, A. J. "Limited-Access Urban Highways." The University of New Mexico. *Proceedings of the Tenth Annual Highway Engineering Conference Held at the University of New Mexico* (1946), 43–50.

Rykken, K. B. "Credit Financing for Highways." *Proceedings of the Thirty-eighth Annual Meeting of the American Association of State Highway Officials* (1952), 139.

Saarinen, Eliel. "Green-Belts, Traffic Efficiency, and Quietness of Living in Urban Areas: A Plea for Purposeful Dreaming in City and Regional Design." *The American City* 58 (April, 1943), 57–59.

St. Clair, G. P. "Bond-Issue Financing of Arterial Highway Improvements." Highway Research Board, *Proceedings of the Twenty-ninth Annual Meeting* (1950), 20–43.

Schulenberg, T. W. "Limited-Access Thoroughfares in the Community Plan." *Traffic Engineering* 25 (September, 1955), 485–489, 493.

Smith, Larry. "Maintaining the Health of Our Central Business Districts." *Traffic Quarterly* 8 (April, 1954), 111–122.

Starick, Herbert W. "Urban Road Problems." The Ohio State University. *Proceedings of the Ohio Highway Engineering Conference.* Experiment Station Bulletin No. 129 (September, 1947), 35–40.

Steele, C. A. "Finding Additional Revenue for Cities." Highway Research Board, *Proceedings of the Twenty-seventh Annual Meeting* 27 (c. 1948), 11–28.

"Superhighways and Subways for Chicago." *Engineering News-Record* 123 (November 23, 1939), 56–58.

Tallamy, B[ertram] D. "Economic Effect of N.Y. Thruway." *Traffic Quarterly* 9 (April, 1955), 220–228.

———. "Meeting the Urban Thoroughfare Challenge." *Proceedings of the Thirty-third Annual Convention of the American Association of State Highway Officials* (1947), 154–159.

"Texas Urban Expressways Being Designed from Center Out." *Roads and Streets* 88 (November, 1945), 65–68.

Tobler, H. Willis. "The Farmer's Need for Good Roads." *Proceedings of the Second Highway Transportation Congress* (1948), 16–17.

"Toll Superhighway System Rejected." *Engineering News-Record* 122 (May 11, 1939), 63–66.

Traer, Will M. "The Florida East Coast Parkway: A Look into the Future." *Florida Public Works* 18 (April, 1941), 3–4, 16.

"Traffic Jams Business Out, Produces Bald Spots in City Centers." *Architectural Forum* 72 (January, 1940), 64–65.

"Trafficway Reclaims a Waterfront." *Engineering News-Record* 124 (May 9, 1940), 68–72.

Tuemmler, Fred W. "Highway Planning in the Maryland-Washington Regional District." *American Road Builders' Association.* Technical Bulletin No. 139 (1948), 5–10.

Tuttle, Lawrence S. "Loads and Roads for War and Peace." *Engineering News-Record* 124 (January 18, 1940), 86–89.

"The U.S. Highway System . . . Is the World's Best Highway System But Is Not a Good Highway System." *Fortune* 23 (June, 1941), 92–99, 106, 108.

Upham, Charles M. "Highway Construction in the Postwar Period." The University of Michigan. *Proceedings of the Thirtieth Annual Highway Conference* (c. 1944), 13–22.

———. "Highways of Tomorrow." *Roads and Streets* 82 (January, 1939), 42–43.

———. "The National Postwar Highway Program." The University of Michigan. *Proceedings of the Twenty-eighth Annual Highway Conference* 43 (May 14, 1942), 35–52.

———. "The Postwar Highway Picture." *Published Proceedings of the Highway Conference.* The University of Tennessee *Record* 46 (July, 1943), 57–62.

———. "The Post-War Highway Program." Purdue University. *Proceedings of the Thirty-second Annual Road School.* Extension Series No. 61 (July, 1946), 15–27.

———. "Road Building Cures Unemployment." *Roads and Streets* 82 (May, 1939), 46–47.

———. "Roads First Need of a Mobile Defense." *Florida Public Works* 17 (September, 1940), 3–4, 8–9.

———. "Safe and Sane Roads." *Roads and Streets* 82 (July, 1939), 34.

"Urban Expressway Progress." *Roads and Streets* 97 (December, 1954), 42–43.

"Urban Expressways Need Flexible Plan." *Roads and Streets* 97 (April, 1954), 68.

Van Tassel, Roger. "Economic Aspects of Expressway Construction." *Journal of the American Institute of Planners* 20 (Spring, 1954), 82–86.

Van Wagoner, Murray D. "Superhighways Ahead." *Proceedings of the Thirty-sixth Annual Convention of the American Road Builders' Association* (c. 1939), 12–20.

———. "Super-Highways and the Farmer." *Farmer's Elevator Guide* 33 (November 5, 1938), 7.

Volk, Wayne N. "An Introduction to City-State Relations in Traffic Matters." *Proceedings of the Thirty-seventh Annual Meeting of the American Association of State Highway Officials* (1951), 418–422.

Wagner, E. F. "Advance Right-of-Way Purchases for Freeway Construction." *Proceedings of the Fortieth Annual Meeting of the American Association of State Highway Officials* (1954), 59–70.

"What Congress Plans to Do on Roads." *Engineering News-Record* 124 (January 18, 1940), 96.

Willey, William E. "Arizona Highway Sufficiency Rating System." *Proceedings of the Thirty-fourth Annual Convention of the American Association of State Highway Officials* (1948), 30–37.

———. "Measurement of Highway Needs by Sufficiency Ratings." *Proceedings of the Thirty-seventh Annual Meeting of the American Association of State Highway Officials* (1951), 22–24.

Williams, Allan M. "Michigan Looks at the National Road Program: The County Viewpoint." The University of Michigan. *Proceedings of the Fortieth Annual Highway Conference* 57 (c. 1955), 11–16.

Williams, Jesse E. "Some Economic Aspects of Highway Construction." The University of New Mexico. *Proceedings of the Tenth Annual Highway Engineering Conference* (1946), 33–42.

Williamson, A. V. "Limited Access Highways." The University of Colorado. *The Seventeenth Annual Highway Conference at the University of Colorado.* Highway Series No. 17 (May, 1944), 5–10.

Winfrey, Robley. "The Need of New Frontiers for Highways." The University of Tennessee *Record* 55 (July, 1952), 26–41.

Yost, Light B. "An Investment in Progress." The University of Tennessee *Record* 58 (July, 1955), 9–12.

Young, J. C. "Economic Effects of Expressways on Business and Land Values." *Traffic Quarterly* 5 (October, 1951), 353–368.

Young, William R. "The Role of the Master Plan in Highway Development." *Civil Engineering* 9 (November, 1939), 651–653.

SECONDARY SOURCES: ARTICLES AND BOOKS

Abrams, Charles. *The City Is the Frontier.* New York, 1965.

Adams, Walter, and Gray, Horace M. *Monopoly in America: The Government as Promoter.* New York, 1955.

Albertson, Dean (ed.). *Eisenhower as President.* New York, 1968.

Alexander, Charles C. *Holding the Line: The Eisenhower Era, 1952–1961.* Bloomington, 1975.

Altshuler, Alan. *The City Planning Process: A Political Analysis.* New York, 1965.

American Association of State Highway Officials. *American Association of State Highway Officials: A Story of the Beginning, Purposes, Growth, Activities, and Achievements of AASHO.* Washington, D.C., 1965.

Arnold, Joseph L. *The New Deal in the Suburbs: A History of the Greenbelt Town Program, 1935–1954.* Columbus, 1971.

Bailey, Stephen K. *Congress Makes a Law: The Story Behind the Employment Act of 1946.* New York, 1950.

Baratz, Morton S. *The American Business System in Transition.* New York, 1970.

Bell, Jack. *The Splendid Misery: The Story of the Presidency and Power Politics at Close Range.* New York, 1960.

Berle, Adolf A. *Power without Property: A New Development in American Political Economy.* New York, 1959.

Bernstein, Irving. *The Lean Years: A History of the American Worker, 1920–1933.* Baltimore, 1966.

Broehl, Wayne G., Jr. *Trucks . . . Trouble . . . and Triumph: The Norwalk Truck Line Company.* New York, 1954.

Brown, A. Theodore. *The Politics of Reform: Kansas City's Municipal Government, 1925–1950.* Kansas City, 1958.

Brownell, Blaine A. "The Commercial-Civic Elite and City Planning in Atlanta, Memphis, and New Orleans in the 1920s." *The Journal of Southern History* 41 (August, 1975), 339–368.

———. "A Symbol of Modernity: Attitudes toward the Automobile in Southern Cities in the 1920s." *American Quarterly* 24 (March, 1972), 20–44.

Burnham, John C. "The Gasoline Tax and the Automobile Revolution." *The*

Mississippi Valley Historical Review: A Journal of American History 98 (December, 1961), 435–459.

Burns, James MacGregor. *Roosevelt: The Lion and the Fox.* New York, 1956.

Caro, Robert A. *The Power Broker: Robert Moses and the Fall of New York.* New York, 1974.

Childs, Marquis. *Eisenhower: Captive Hero. A Critical Study of the General and the President.* New York, 1958.

Condit, Carl W. *Chicago, 1910–1929: Building, Planning, and Urban Technology.* Chicago, 1973.

———. *Chicago, 1930–1970: Building, Planning, and Urban Technology.* Chicago, 1974.

Dahl, Robert A. *Who Governs? Democracy and Power in an American City.* New Haven, 1961.

Dale, Edwin L., Jr. *Conservatives in Power: A Study in Frustration.* Garden City, New York, 1960.

Dalfiume, Richard M. (ed.). *American Politics since 1945.* Chicago, 1969.

Davies, Richard O. (author and ed.). *The Age of Asphalt: The Automobile, the Freeway, and the Condition of Metropolitan America.* Philadelphia, 1975.

———. *Housing Reform during the Truman Administration.* Columbia, Missouri, 1966.

Derber, Milton. "The New Deal and Labor," pp. 110–132 in Braeman, John, et al. (eds.). *The New Deal: The National Level.* Columbus, 1975.

Eulau, Heinz. *Class and Party in the Eisenhower Years: Class Roles and Perspectives in the 1952 and 1956 Elections.* New York, 1962.

Flink, James J. *America Adopts the Automobile, 1895–1910.* Cambridge, 1970.

———. *The Car Culture.* Cambridge, 1975.

———. "Three Stages of American Automobile Consciousness." *American Quarterly* 24 (October, 1972), 451–473.

Foster, Mark S. "The Model-T, the Hard Sell, and Los Angeles's Urban Growth: The Decentralization of Los Angeles during the 1920s." *Pacific Historical Review* 44 (November, 1975), 459–484.

Galbraith, John Kenneth. *American Capitalism: The Concept of Countervailing Power.* Boston, 1956.

Gans, Herbert J. *People and Plans: Essays on Urban Problems and Solutions.* New York, 1968.

Goldman, Eric. *The Crucial Decade—And After: America, 1945–1960.* New York, 1960.

Gordon, Mitchell. *Sick Cities.* Baltimore, 1965.

Gottman, Jean. *Megalopolis: The Urbanized Northeastern Seaboard of the United States.* New York, 1961.

Graham, Otis L., Jr. "The Planning Ideal and American Reality: The 1930s," pp. 257–299 in Elkins, Stanley, and McKitrick, Eric (eds.). *The Hofstadter Aegis: A Memorial.* New York, 1974.

———. *Toward a Planned Society: From Roosevelt to Nixon.* New York, 1976.

Green, Constance M. *The Rise of Urban America.* New York, 1965.

Green, Scott. *The Emerging City: Myth and Reality.* New York, 1962.

Harris, Louis. *Is There a Republican Majority? Political Trends, 1952–1956.* New York, 1954.

Harris, Seymour E. *The Economics of the Political Parties with Special Attention to Presidents Eisenhower and Kennedy.* New York, 1962.

Hawley, Ellis W. "The New Deal and Business," pp. 50–82 in Braeman, John, et al. (eds.). *The New Deal: The National Level.* Columbus, 1975.

———. *The New Deal and the Problem of Monopoly: A Study in Economic Ambivalence.* Princeton, New Jersey, 1966.

Hays, Samuel P. "New Possibilities for American Political History: The Social Analysis of Political Life," pp. 181–227 in Lipset, Seymour M., and Hofstadter, Richard (eds.). *Sociology and History: Methods.* New York, 1968.

———. "Political Parties and the Community-Society Continuum," pp. 152–181 in Chambers, William N., and Burnham, Walter Dean (eds.). *The American Party Systems: Stages of Political Development.* New York, 1967.

Hebden, Norman, and Smith, Wilbur S. *State-City Relationships in Highway Affairs.* New Haven, 1950.

Holt, Glen E. "The Changing Perception of Urban Pathology: An Essay on the Development of Mass Transit in the United States," pp. 324–343 in Jackson, Kenneth T., and Schultz, Stanley K. (eds.). *Cities in American History.* New York, 1972.

Holt, W. Stull. *The Bureau of Public Roads: Its History, Activities and Organization.* Baltimore, 1923.

Horwood, Edgar M., et al. *Studies of the Central Business District and Urban Freeway Development.* Seattle, 1959.

Kaplan, Harold. *Urban Renewal Politics: Slum Clearance in Newark.* New York, 1963.

Kelley, Ben. *The Pavers and the Paved.* New York, 1971.

Key, V. O., Jr. *The Responsible Electorate: Rationality in Presidential Voting, 1936–1960.* New York, 1968.

Kolko, Gabriel. *Wealth and Power in America: An Analysis of Social Class and Income Distribution.* New York, 1968.

Leavitt, Helen. *Superhighway-Superhoax.* Garden City, New York, 1970.

Leuchtenburg, William E. *Franklin D. Roosevelt and the New Deal, 1932–1940.* New York, 1963.

Link, Arthur S., et al. *American Epoch: A History of the United States since the 1890s.* Volume II: *1921–1941*; Volume III: *1938–1966.* New York, 1967.

Lipset, Seymour M. *Political Man: The Social Bases of Politics.* Garden City, New York, 1963.

Lowe, Jeanne R. *Cities in a Race—With Time: Progress and Poverty in America's Renewing Cities.* New York, 1967.

Lubell, Samuel. *Revolt of the Moderates.* New York, 1956.

Lubove, Roy. *Twentieth-Century Pittsburgh: Government, Business, and Environmental Change.* New York, 1969.

Lynd, Robert S., and Lynd, Helen M. *Middletown: A Study in American Culture.* New York, 1956.

———. *Middletown in Transition: A Study in Cultural Conflicts.* New York, 1937.

McCoy, Donald R. *Coming of Age: The United States during the 1920s and 1930s.* Baltimore, 1973.

McKelvey, Blake. *The Emergence of Metropolitan America, 1915–1916.* New Brunswick, New Jersey, 1968.

Madgwick, P. J. "The Politics of Urban Renewal." *Journal of American Studies* 5 (December, 1971), 265–280.
Mayer, George H. *The Republican Party, 1854–1966*. New York, 1967.
Mills, C. Wright. *The Power Elite*. New York, 1956.
Moline, Norman T. *Mobility and the Small Town, 1900–1930: Transportation Change in Oregon, Illinois*. Chicago, 1971.
Mooney, Booth. *The Politicians: 1945–1960*. New York, 1970.
Morehouse, Thomas A. "Artful Interpretation: The 1962 Highway Act," pp. 353–368 in Danielson, Michael N. (ed.). *Metropolitan Politics: A Reader*. 2d Edition. Boston, 1971.
Mowbray, A. Q. *Road to Ruin*. Philadelphia, 1968.
Mowry, George E. *The Urban Nation: 1920–1960*. New York, 1965.
Nash, Gerald D. *United States Oil Policy 1890–1964: Business and Government in Twentieth-Century America*. Pittsburgh, 1968.
Nossiter, Bernard D. *The Mythmakers: An Essay on Power and Wealth*. Boston, 1967.
O'Neill, William L. (ed.). *American Society Since 1945*. Chicago, 1969.
Owen, Wilfred. *Cities in the Motor Age*. New York, 1959.
Patterson, James T. *Congressional Conservatism and the New Deal: The Growth of the Conservative Coalition in Congress, 1933–1939*. Lexington, 1967.
———. *The New Deal and the States: Federalism in Transition*. Princeton, New Jersey, 1969.
Paxson, Frederic L. "The American Highway Movement, 1916–1935." *American Historical Review* 101 (January, 1946), 236–253.
Peterson, Jon A. "The City Beautiful Movement: Forgotten Origins and Lost Meanings." *Journal of Urban History* 2 (August, 1976), 415–434.
Polenberg, Richard. "The Decline of the New Deal," pp. 246–266 in Braeman, John, et al. (eds.). *The New Deal: The National Level*. Columbus, 1975.
———. *War and Society: The United States, 1941–1945*. Philadelphia, 1972.
Polsby, Nelson W. *Community Power and Political Theory*. New Haven, 1963.
Powledge, Fred. *Model City, a Test of American Liberalism: One Town's Efforts to Rebuild Itself*. New York, 1970.
Pusey, Merlo J. *Eisenhower the President*. New York, 1956.
Rabinowitz, Francine F. *City Politics and Planning*. New York, 1969.
Rae, John B. *The American Automobile: A Brief History*. Chicago, 1965.
———. "Coleman du Pont and His Road." *Delaware History* 16 (Spring-Summer 1975), 171–183.
———. *The Road and the Car in American Life*. Cambridge, 1971.
Reichard, Gary W. *The Reaffirmation of Republicanism: Eisenhower and the Eighty-third Congress*. Knoxville, 1975.
Reps, John W. *The Making of Urban America: A History of City Planning in the United States*. Princeton, New Jersey, 1965.
Richardson, Elmo. *Dams, Parks and Politics: Resource Development and Preservation in the Truman-Eisenhower Era*. Lexington, 1973.
Rose, Mark H. "PAR—Project Adequate Roads: Traffic Jams, Business, and Government, 1945–1956," pp. 120–136 in Uselding, Paul (ed.). *Business and Economic History*. Urbana, 1975.
———. "The Politics of Highways," pp. 64–68 in Davies, Richard O. (author

and ed.). *The Age of Asphalt: The Automobile, the Freeway, and the Condition of Metropolitan America.* Philadelphia, 1975.
Rovere, Richard H. *Affairs of State: The Eisenhower Years.* New York, 1956.
Schwartz, Gary T. "Urban Freeways and the Interstate System." *Southern California Law Review* 49 (March, 1976), 406–513.
Scott, Mel. *American City Planning since 1890: A History Commemorating the Fiftieth Anniversary of the American Institute of Planners.* Berkeley, 1969.
Smerk, George M. *Urban Transportation: The Federal Role.* Bloomington, 1965.
Sternsher, Bernard. *Rexford Tugwell and the New Deal.* New Brunswick, New Jersey, 1964.
Stewart, Murray (ed.). *The City: Problems of Planning.* Baltimore, 1972.
Taff, Charles A. *Commercial Motor Transportation.* Homewood, Illinois, 1955.
Talbot, Allan R. *The Mayor's Game: Richard Lee of New Haven and the Politics of Change.* New York, 1967.
Van Metre, T. W. *Transportation in the United States.* Chicago, 1939.
Vatter, Harold G. *The U.S. Economy in the 1950s: An Economic History.* New York, 1963.
Wardlow, Chester. *United States Army World War II, The Transportation Corps: Responsibilities, Organization, and Operations.* Washington, D.C., 1951.
Warner, Sam B., Jr. *The Private City: Philadelphia in Three Periods of Its Growth.* Philadelphia, 1968.
———. *Streetcar Suburbs: The Process of Growth in Boston, 1870–1900.* New York, 1970.
———. *The Urban Wilderness: A History of the American City.* New York, 1972.
Wiebe, Robert H. *The Search for Order, 1870–1920.* New York, 1967.
Williams, William A. *The Contours of American History.* Cleveland, 1961.
Wilson, James Q. (ed.). *Urban Renewal: The Record and the Controversy.* Cambridge, 1968.
Wilson, William H. *The City Beautiful Movement in Kansas City.* Columbia, Missouri, 1964.

UNPUBLISHED STUDIES

Adams, Robert W. "Urban Renewal Politics: A Case Study of Columbus, Ohio, 1952–1961." Ph.D. dissertation, The Ohio State University, 1970.
Annunziata, Frank. "The Attack on the Welfare State: Patterns of Anti-Statism from the New Deal to the New Left." Ph.D. dissertation, The Ohio State University, 1968.
Belasco, Warren J. "Americans on the Road: Autocamping, Tourist Camps, Motels, 1910–1945." Ph.D. dissertation, The University of Michigan, 1977.
Brownell, Blaine A. "The Automobile and Urban Planning in the 1920s: The Cases of Three Cities in the American South." Paper presented at the meetings of the Organization of American Historians, Denver, Colorado, April 19, 1974.
———. "The Automobile and Urban Structure." Paper presented at the meet-

ings of the American Studies Association, San Antonio, Texas, November 6, 1975.

Campbell, Warren M. "Politics, Procedures, and City Planning: The Master Plan as an Instrument of Policy Formulation and Control." Ph.D. dissertation, Stanford University, 1963.

Crane, Stuart. "Federal Financing for Toll Projects Incorporated in the Interstate Highway System." D.B.A. dissertation, Indiana University, 1967.

Flink, James J. "Automobility and the National Parks." Paper presented at the meetings of the American Studies Association, San Antonio, Texas, November 6, 1975.

———. "Mass Automobility: An Urban Reform that Backfired." Paper presented at the meetings of the Missouri Valley History Conference, Omaha, Nebraska, March 6, 1975.

Goodwin, Herbert M. "California's Growing Freeway System." Ph.D. dissertation, The University of California, Los Angeles, 1969.

Johnston, Norman J. "Harland Bartholomew: His Comprehensive Plans and Science of Planning." Ph.D. dissertation, The University of Pennsylvania, 1964.

Lind, William E. "Thomas H. MacDonald: A Study of the Career of an Engineer and His Influence on Public Roads in the United States, 1919–1953." M.A. thesis, The American University, 1965.

Loretz, John. "John Nolen and the American City Planning Movement: A History of Culture Change and Community Response, 1900–1940." Ph.D. dissertation, The University of Pennsylvania, 1964.

Peters, Kenneth E. "The Good Roads Movement and the Michigan State Highway Department, 1905–1919." Ph.D. dissertation, University of Michigan, 1972.

———. "Michigan Good Roads Politics, 1900–1917." Paper presented at the meetings of the Missouri Valley History Conference, Omaha, Nebraska, March 12, 1976.

Riddick, Winston W. "The Politics of National Highway Policy, 1953–1966." Ph.D. dissertation, Columbia University, 1973.

Saulnier, Raymond J. "Anti-Inflation Policies in President Eisenhower's Second Term." Paper presented at the meetings of the American Historical Association, San Francisco, California, December 30, 1973.

Sponholtz, Lloyd L. "The Good Roads Movement in Ohio, 1900–1912." Paper presented at the meetings of the Missouri Valley History Conference, Omaha, Nebraska, March 12, 1976.

Van Barrow, Robert. "The Politics of Interstate Route Selection: A Case Study of Interest Activities in a Decision Situation." Ph.D. dissertation, Florida State University, 1967.

OTHER SOURCES

Bresnahan, William A. President of the American Trucking Associations, Incorporated. Personal interview, June 24, 1971.

Creighton, Richard C. Director, Highway Division, Associated General Contractors of America, Incorporated. Personal interview, June 23, 1971.

Selected Bibliography

Turner, Francis C. Federal Highway Administrator. Personal interview, July 26, 1971.

Index